Advanced Structured Materials

Volume 95

Common engineering materials reach in many applications their limits and new developments are required to fulfil increasing demands on engineering materials. The performance of materials can be increased by combining different materials to achieve better properties than a single constituent or by shaping the material or constituents in a specific structure. The interaction between material and structure may arise on different length scales, such as micro-, meso- or macroscale, and offers possible applications in quite diverse fields.

This book series addresses the fundamental relationship between materials and their structure on the overall properties (e.g. mechanical, thermal, chemical or magnetic etc.) and applications.

The topics of *Advanced Structured Materials* include but are not limited to

- classical fibre-reinforced composites (e.g. class, carbon or Aramid reinforced plastics)
- metal matrix composites (MMCs)
- micro porous composites
- micro channel materials
- multilayered materials
- cellular materials (e.g. metallic or polymer foams, sponges, hollow sphere structures)
- porous materials
- truss structures
- nanocomposite materials
- biomaterials
- nano porous metals
- concrete
- coated materials
- smart materials

Advanced Structures Material is indexed in Google Scholar and Scopus.

More information about this series at http://www.springer.com/series/8611

Heiko Herrmann · Jürgen Schnell
Editors

Short Fibre Reinforced Cementitious Composites and Ceramics

Springer

Editors
Heiko Herrmann
School of Science, Department of Cybernetics
Tallinn University of Technology
Tallinn, Estonia

Jürgen Schnell
Fachbereich Bauingenieurwesen
Technische Universität Kaiserslautern
Kaiserslautern, Germany

ISSN 1869-8433 ISSN 1869-8441 (electronic)
Advanced Structured Materials
ISBN 978-3-030-00867-3 ISBN 978-3-030-00868-0 (eBook)
https://doi.org/10.1007/978-3-030-00868-0

Library of Congress Control Number: 2018963028

This Springer imprint is published by the registered company Springer Nature Switzerland AG
The registered company address is: Gewerbestrasse 11, 6330 Cham, Switzerland

Preface

Composites containing short fibers are important in many technological fields, and the used composites range from fiber plastics to fiber concretes. Especially in the building industry, fiber composites are gaining importance, as they may increase the building speed and improve material properties. The used fibers include steel, basalt, carbon, and polymer fibers, of many different shapes and aspect ratios. The target of including fibers also differs, including improving tensile strength, post-cracking behavior (ductility), and temperature resistance. In all cases, the microstructure and spatial and orientational distribution of the fibers are important. At this colloquium, different methods and results for the analysis of microstructure and fiber orientation analysis in cementitious building materials containing fibers, among them concretes and refractory composites, have been presented. The goal of the colloquium was to present a platform for the exchange of ideas between different fields with similar problems. While having different chemical compositions and length scales, cementitious composited and ceramics reinforced with short fibers share similar mechanical properties and theoretical problems. Especially, the addition of the short fibers makes the composite stochastically anisotropic and inhomogeneous, since several macro- and mesoscale properties depend on the spatial and orientational distribution of fibers. An important aspect is the microstructure around the fibers and the adhesion of the matrix to the fibers, which has large influence on mechanical properties. There were altogether 19 participants and 14 presentations, among these three keynote talks, given by Daniele Casucci (Kaiserslautern), Johan L. Silfwerbrand (Stockholm), and Akke Suiker (Eindhoven). This volume contains selected papers from participants and members of the scientific committee.

We thank Euromech for accepting the proposal of this colloquium and for supporting it, both financially and by hosting the website. Support by the Estonian Research Council's exploratory research grant PUT1146 is gratefully acknowledged. Furthermore, we thank Holm Altenbach (as series editor) and Christoph

Baumann (Springer) for supporting this book proposal. For their help with organizational matters, we thank Kristi Juske, Mare Saago, Oksana Goidyk, and Andres Braunbrück. Finally, we thank TUT Mektory for providing the rooms for the colloquium.

Tallinn, Estonia
Kaiserslautern, Germany
June 2018

Heiko Herrmann
Jürgen Schnell

Contents

Contributors

Andres Braunbrück Department of Cybernetics, Department of Civil Engineering and Architecture, Tallinn University of Technology, Tallinn, Estonia

Daniene Casucci Institute of Concrete Structures and Structural Engineering, Technische Universität Kaiserslautern, Kaiserslautern, Germany; Hilti Corporation, Schaan, Liechtenstein

Oksana Goidyk Department of Cybernetics, Tallinn University of Technology, Tallinn, Estonia

Dorys C. González Department of Civil Engineering, University of Burgos, Burgos, Spain

Heiko Herrmann Department of Cybernetics, Tallinn University of Technology, Tallinn, Estonia

Irina Hussainova Tallinn University of Technology, Tallinn, Estonia

Roman Ivanov Tallinn University of Technology, Tallinn, Estonia

Iwona Jasiuk University of Illinois at Urbana-Champaign, Urbana, IL, USA

S. Sudhir Kale University of Illinois at Urbana-Champaign, Urbana, IL, USA

Jacek Katzer Koszalin University of Technology, Koszalin, Poland

Olga Kononova Concrete Mechanics laboratory, Institute of Mechanics, Riga Technical University, Riga, Latvia

Andrejs Krasnikovs Concrete Mechanics laboratory, Institute of Mechanics, Riga Technical University, Riga, Latvia

Videvuds Lapsa Faculty of Civil Engineering, Institute of Building Production, Riga Technical University, Riga, Latvia

Arturs Lukasenoks Concrete Mechanics laboratory, Institute of Mechanics, Riga Technical University, Riga, Latvia

Arturs Macanovskis Concrete Mechanics laboratory, Institute of Mechanics, Riga Technical University, Riga, Latvia

Jesús Mínguez Department of Civil Engineering, University of Burgos, Burgos, Spain

Hendrik Naar Department of Civil Engineering and Architecture, Tallinn University of Technology, Tallinn, Estonia

Tyler Oesch Bundesanstalt für Materialforschung und –prüfung, Federal Institute for Materials Research and Testing, Berlin, Germany

Tomasz Ponikiewski Silesian University of Technology, Gliwice, Poland

Gonzalo Ruiz Department of Applied Mechanics, University of Castilla – La Mancha, Ciudad Real, Spain

Jürgen Schnell Institute of Concrete Structures and Structural Engineering, Technische Universität Kaiserslautern, Kaiserslautern, Germany

Ludwig Stelzner Bundesanstalt für Materialforschung und –prüfung, Federal Institute for Materials Research and Testing, Berlin, Germany

Manuel Tarifa Department of Applied Mechanics, University of Castilla – La Mancha, Ciudad Real, Spain

Catherina Thiele Institute of Concrete Structures and Structural Engineering, Technische Universität Kaiserslautern, Kaiserslautern, Germany

Tanel Tuisk Department of Civil Engineering and Architecture, Tallinn University of Technology, Tallinn, Estonia

Miguel A. Vicente Department of Civil Engineering, University of Burgos, Burgos, Spain

Frank Weise Bundesanstalt für Materialforschung und –prüfung, Federal Institute for Materials Research and Testing, Berlin, Germany

Xiaoxing Zhang Department of Applied Mechanics, University of Castilla – La Mancha, Ciudad Real, Spain

Study of Crack Patterns of Fiber-Reinforced Concrete (FRC) Specimens Subjected to Static and Fatigue Testings Using CT-Scan Technology

Miguel A. Vicente, Gonzalo Ruiz, Dorys C. González, Jesús Mínguez, Manuel Tarifa and Xiaoxing Zhang

Abstract This paper demonstrates the widely accepted hypothesis that the compressive testing is a particular case of a cyclic test where failure occurs during the first cycle. To perform this, a test on 32 fiber-reinforced high-performance concrete specimens have been carried out. Sixteen of them have been tested under low-cycle fatigue compressive loading up to failure. Eight of them have been tested under monotonic compressive loading, until failure too. And the last eight specimens have remained intact. All of them have been scanned using a Computed Tomography (CT) Scan in order to define the pattern of their damage, which includes voids and cracks. The results show that the average damage maps of monotonic and fatigue series are statistically identical, which confirms the hypothesis previously described. In addition, both series are different to the intact series, which means that not a random damage distribution occurs when specimens collapse.

M. A. Vicente (✉) · D. C. González · J. Mínguez
Department of Civil Engineering, University of Burgos, Burgos, Spain
e-mail: mvicente@ubu.es

D. C. González
e-mail: dgonzalez@ubu.es

J. Mínguez
e-mail: jminguez@ubu.es

G. Ruiz · M. Tarifa · X. Zhang
Department of Applied Mechanics, University of Castilla – La Mancha, Ciudad Real, Spain
e-mail: Gonzalo.Ruiz@uclm.es

M. Tarifa
e-mail: ManuelAgustin.Tarifa@uclm.es

X. Zhang
e-mail: zhangxiaoxinhrb@gmail.com

H. Herrmann and J. Schnell (eds.), *Short Fibre Reinforced Cementitious Composites and Ceramics*, Advanced Structured Materials 95,
https://doi.org/10.1007/978-3-030-00868-0_1

1 Introduction

During the last years, the progressive increase of the strength of concrete is leading to slender structures, where cyclic loads have a more relevant influence. Thus, fatigue phenomena in concrete are of increasing interest and the development of accurate predictive models has a great scientific interest.

A significant amount of fatigue models for concrete have been developed until now. They establish a relationship between the maximum and the minimum stress ratio of the cyclic load with the number of cycles up to failure (usually called as "fatigue life") [1, 4, 7, 11, 12, 15, 18, 19].

Most fatigue models consider the hypothesis of convergence to the "initial distribution", which means that the static testing is a particular case of cyclic testing where the fatigue life is equal to 1. This hypothesis is supported by the fact that the crack patterns obtained in both cases are equal.

The aim of the research conducted by the authors is to check if this hypothesis is true or false. To perform this, the Computed Tomography (CT) technology is used, in combination with a homemade post-processing software, in order to define the crack pattern within the specimens.

Computed tomography is a nondestructive technique, based on absorbing X-rays, that permits the visualization of the internal microstructure of material up to microrange resolution.

The field of application is very wide. This is a well-known technology in medicine, because of its enormous advantages, but it is also very useful in other fields. For example, this technology is also very common in veterinarian, or paleontology. In materials engineering, this technology is starting to be widely used to analyse the internal microstructure of a wide variety of materials: metals, rocks, composites, etc.

This technology is also very useful for the study of concrete microstructure. Most of the macroscopic responses of concrete elements can be explained through the understanding of the microstructure. For example, the freeze-thaw behavior of concrete strongly depends on pore sizes and pore distribution. Moreover, in case of fiber-reinforced concrete, the use of CT-Scan can provide useful information related to the fiber distribution and orientation, which strongly influences its macroscopic response [5, 9, 10, 14, 16].

In recent years, many research have been conducted in order to study the internal microstructure of concrete using this technology [2, 3, 5, 6, 8–10, 13, 14, 17].

The research shown in this paper uses the CT-Scan or detect internal "damage" of concrete, which includes pores and cracks. This technology is able to define the density of each specimen voxel by assigning a shade of a grey. Light shades of grey correspond to high density and dark shades of grey correspond to low density. A more detailed explanation of the CT-Scan technology can be found in [17].

Table 1 Specimens tested under cyclic load (cyclic series)

Cycles for $\sigma_{max} = 0.82 \cdot f_c$; $\sigma_{min} = 0.36 \cdot f_c$							
Specimen	Cycles	Specimen	Cycles	Specimen	Cycles	Specimen	Cycles
1f	1066	5f	1115686[a]	9f	19186	13f	2700
2f	3128	6f	25	10f	90	14f	88
3f	214	7f	111	11f	144	15f	152076
4f	29189	8f	460	12f	3305	16f	1

[a]Run-out

2 Experimental Procedure

Next, the experimental procedure and the scanning procedure are described.

2.1 *Materials Characterization*

In this case, fiber reinforced high performance concrete cubic specimens were performed. Their shape was cubic with 40 mm in edge-length, cut from prism of 150 × 150 × 700 mm.

The concrete compressive strength f_c was 101 MPa, with a standard deviation of 3 MPa.

A total of three series of concrete cubes were performed, named "intact", "monotonic" and "cyclic". The intact and the monotonic series were composed by 8 specimens each. The cyclic series was composed by 16 specimens. A total of 32 specimens were performed.

Specimens belonging to monotonic series were subjected to a monotonic compressive load up to failure. Specimens belonging to cyclic series were subjected to a low-cycle cyclic load up to failure. Finally, specimens belonging to intact series were not tested.

2.2 *Fatigue Tests*

Specimens belonging to cyclic series were subjected to low-cycle cyclic load up to failure. The tests were carried out at a loading frequency of 10 Hz under sinusoidal stress cycles, varying between $0.36 \cdot f_c$ and $0.82 \cdot f_c$. Table 1 shows the fatigue life of all the specimens.

The tests were carried out using a hydraulic jack with a load capacity of ±250 kN.

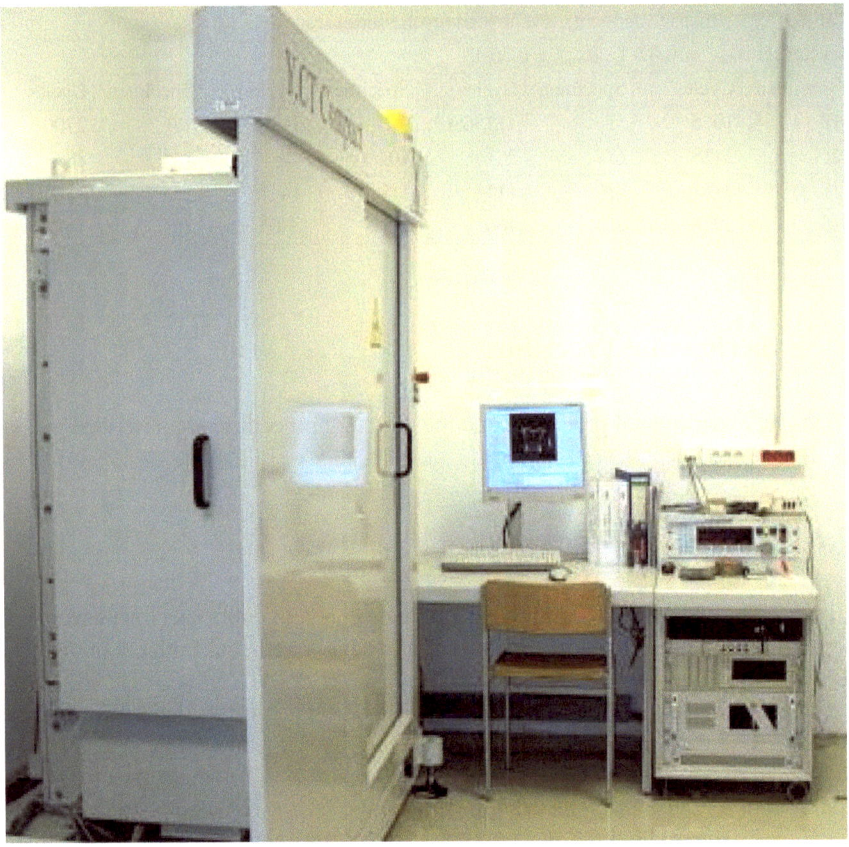

Fig. 1 CT-Scan: Y.CT COMPACT with a YXlon tube of 225 kV/30mA at University of Burgos (Spain)

The results show a wide scatter in the service life of the test specimens. This is very common in concrete and it is due to the inherent scatter of the material, which particularly conducts to a wide scatter in terms of service life.

2.3 *Scanning of the Specimens*

Once the specimens belonging to the monotonic and the cyclic series had been tested, they were scanned using a CT-Scan. In addition, the specimens belonging to the intact series were also scanned.

The CT-Scan used was a Y.CT COMPACT device of the University of Burgos (Spain). It is equipped with a YXlon tube of 225 kV/30mA (Fig. 1). The CT-Scan has a post-processing software which provides 2D slices of 1024 $\times$ 1024 pixels. Thus,

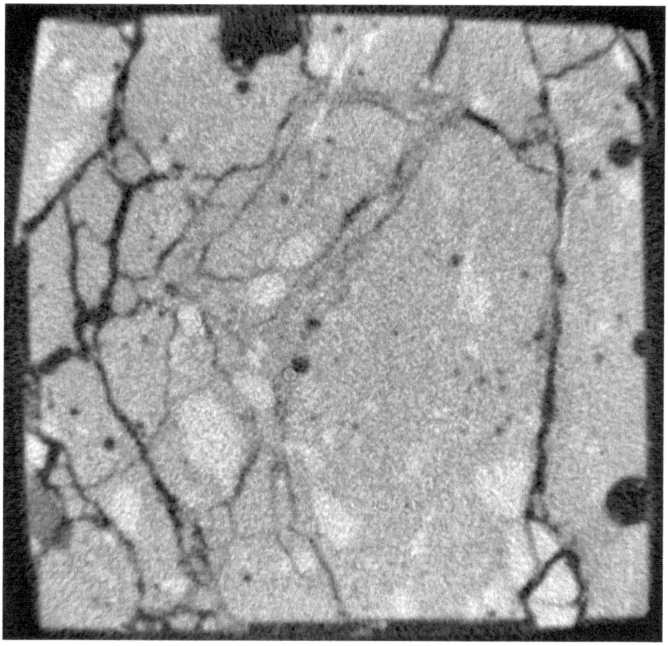

Fig. 2 CT-Scan slide

for a section of $40 \times 40\ \text{mm}^2$, the resolution of the scanner is $55 \times 55\ \mu\text{m}^2$. The vertical distance between slices is 100 μm. The amount of slices per specimen is 401. The voxel has a volume of $55 \times 55 \times 100\ \mu\text{m}^3$.

Each voxel is identified by its center of gravity (coordinates X, Y and Z) and a grey color belonging to a grey scale, from black to white depending on the voxel density. A total amount of 256 grey colors are identified in the grey scale.

Light grey corresponds to high density and dark grey corresponds to low density. Figure 2 shows a slice of a specimen. The total number of voxel is approximately $4.2 \cdot 10^8$.

The post-processing methodology is as follows.

First, once all the voxels had been identified, the ones belonging to damage were extracted. These "damaged" voxels (or empty voxels) are the ones with a density (i.e. a grey color) below a threshold. The result is a 3D image containing only the empty voxels, i.e. the voxels belonging to pores or cracks (Fig. 3).

The average number of empty voxels per specimen in every series is shown in Table 2. In case of intact series, empty voxels belongs only to voids, meanwhile in case of monotonic and cyclic series, empty voxels belongs to voids and cracks.

It should be noted that the number of empty voxels belonging to the intact series are significantly smaller than the ones belonging to monotonic and cyclic series. That is because empty voxels in intact series belong to pores only, meanwhile empty voxels in monotonic and cyclic series includes pores and cracks.

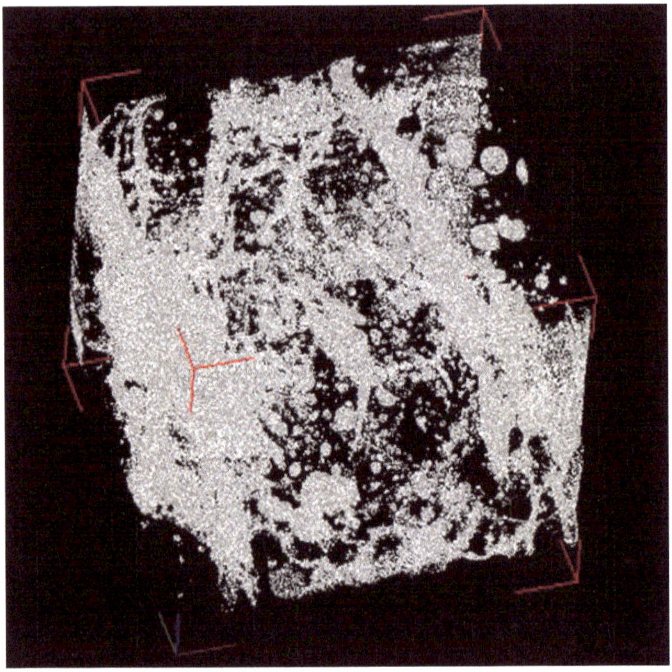

Fig. 3 Example of empty voxels distribution

Table 2 Empty voxels per specimen in every series

Series		Intact	Monotonic	Cyclic
Number of Specimens		8	8	16
Voxels	Mean	73,254	1,472,215	1,089,281
	Std. Dev.	26,134	238,492	438,796
% Voxels	Mean	0.02%	0.35%	0.26%
	Std. Dev.	0.01%	0.06%	0.10%

In order to compare the empty voxel distributions (crack patterns) from the different series, a novel numerical procedure, designed by the authors, have been developed, named "Circumferential test". Next, this procedure is explained.

2.4 *Circumferential Test*

The circumferential test is the procedure for analyzing the raw data from the CT-scan in order to disclose the extent of damage generated in the mechanical tests.

Next, the steps for analyzing the data are exposed:

First, the coordinates x_i, y_i and z_i of each individual voxel are normalized according to the following expression:

$$x_{rel,i} = \frac{2 \cdot x_i}{x_{max}} \tag{1a}$$

$$y_{rel,i} = \frac{2 \cdot y_i}{y_{max}} \tag{1b}$$

$$z_{rel,i} = \frac{z_i}{z_{max}} \tag{1c}$$

where x_{max}, y_{max} and z_{max} are the real dimensions of each individual specimen.

Next, each voxel is identified by a pair of coordinates, the normalized distance of the voxel to the center of gravity of the cross-section where the voxel is placed d_i and the height of the voxel h_i, according to the following expressions (Fig. 4):

$$d_i = \sqrt{\left(x_{rel,i} - x_{G,rel,i}\right)^2 + \left(y_{rel,i} - y_{G,rel,i}\right)^2} \tag{2}$$

$$h_i = z_{rel,i} \tag{3}$$

where $x_{G,rel,i}$ and $y_{G,rel,i}$ are the normalized coordinates of the center of gravity of the considered section. Coordinate d_i varies from 0 to $\sqrt{2}$ while h_i varies from 0 to 1.

Coordinates d and h are divided into twenty subdivisions each, so that the voxels are clustered for all the combinations of d and h coordinates. Thus, the whole cube volume is divided into 400 sub-volumes. For each, the relative frequency of occurrence of empty voxels is calculated, according to the following expression:

$$\text{Relative frequency, } i = \frac{N_i}{N_t} \tag{4}$$

where N_i is the number of empty voxels belonging to each individual sub-volume and N_t is the total number of empty voxel of the specimen.

In addition, a random damage distribution was simulated inside a theoretical 40 mm edge-length using a Monte Carlo model with 10^6 points. This theoretical distribution was used to compare it with the real distribution in order to decide if empty voxels in real specimens are randomly distributed or not.

3 Results and Discussion

Next, the results of CT-Scan are shown. First, the results are shown by means of 3D histograms of each specimen. In addition, the histogram of the random distribution is shown. Second, the results are compressed and shown by means of 2D histograms in the d direction first and in the h directions next, in order to be analyzed appropriately.

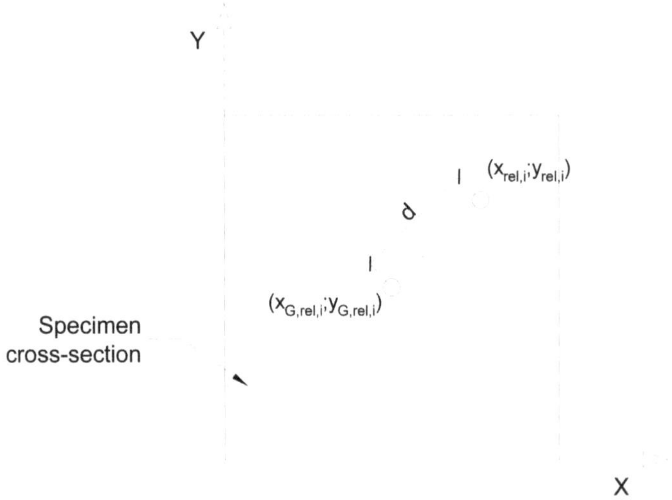

Fig. 4 Voxel coordinates

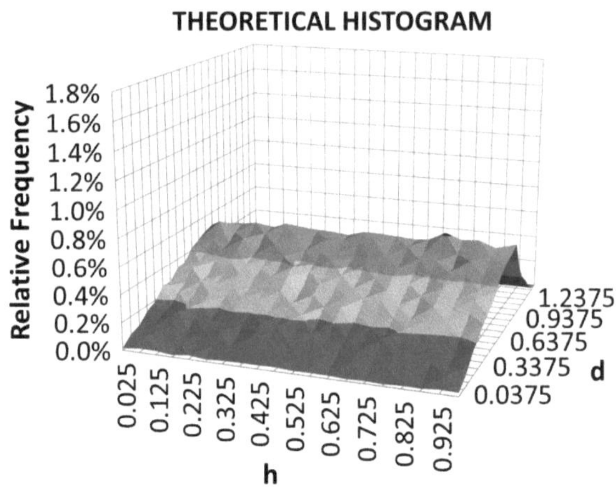

Fig. 5 Theoretical 3D histogram

3.1 3D Histograms

Using all the information explained above, a 3D histogram can be drawn, where x-axis is the parameter h, y-axis is the parameter d and z-axis is the relative frequency. Figures 5, 6, 7 and 8 show all the 3D histograms.

Note than random histogram shows a triangular shape along d parameter, with a maximum at $d = 1$, since such sub-volume corresponds to the longest ring inside the cube.

Histograms belonging to intact series specimens show high peaks. These ones represent the presence of large pores with a great amount of empty voxels.

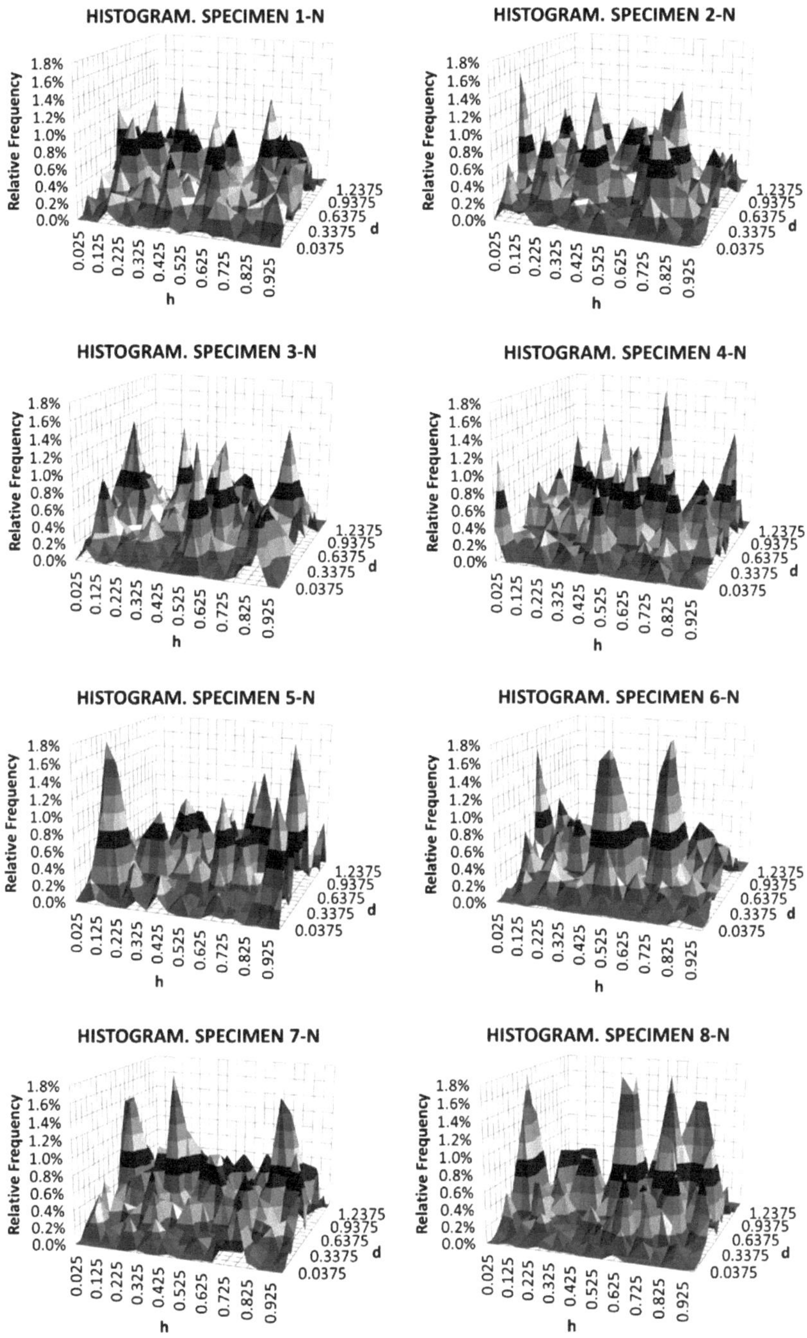

Fig. 6 3D histogram of the intact seriesspecimens

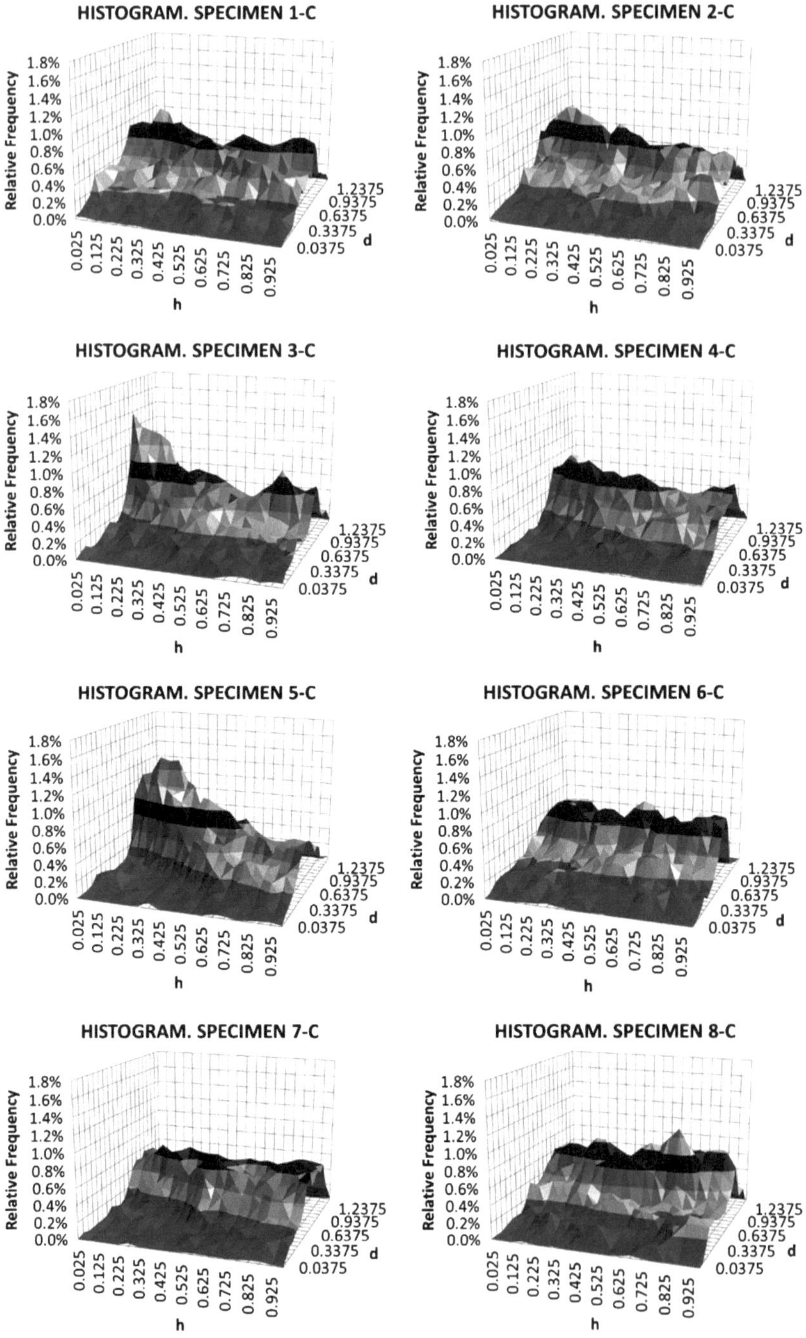

Fig. 7 3D histogram of the monotonic series specimens

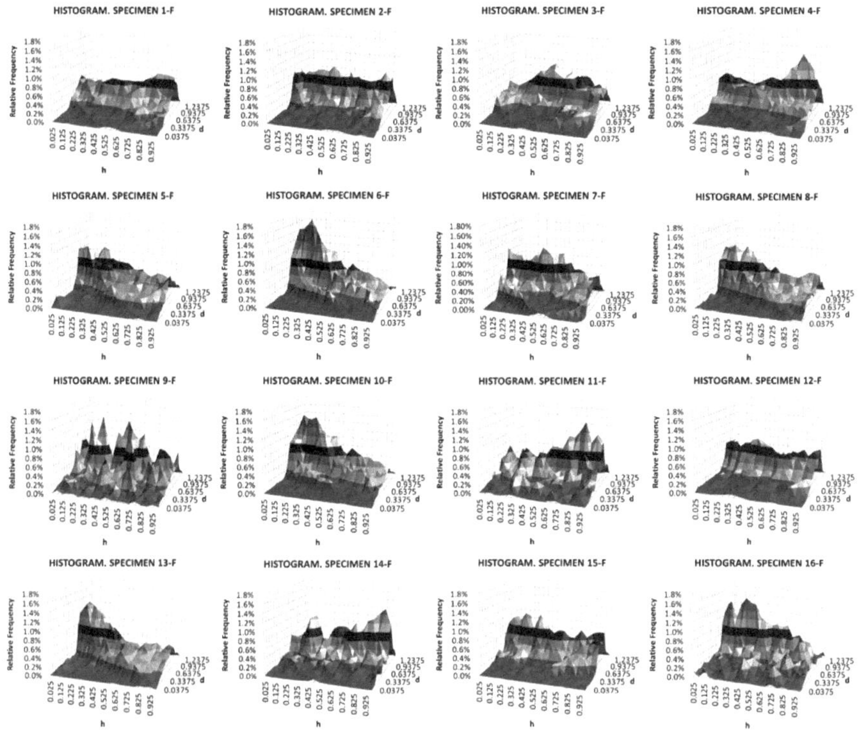

Fig. 8 3D histogram of the cyclic series specimens

In case of monotonic and cyclic series specimens, less peaks are observed. That is because of the presence of cracks, which show more distributed empty pores along the specimen. In consequence, the presence of pores is mitigated.

3.2 2D Histograms Along the H Direction

In order to easily analyze the results, the 3D histograms were aggregated along the h direction, obtaining 2D histograms which represent the relative frequency of occurrence of empty voxels along the d coordinate, i.e., transverse to the load direction. Figures 9, 10 and 11 show the 2D histograms of the different series. Each histogram show the individual data of the specimens (drawn as bar diagrams) and also the average histogram. Additionally, the histogram belonging to the random distribution is shown, in order to be able to compare real and theoretical distributions.

Figures 11, 12 and 13 show the comparison between the theoretical distribution and the real distributions of intact, monotonic and cyclic series. Additionally, the 90% confidence interval is defined by the upper and the lower limits.

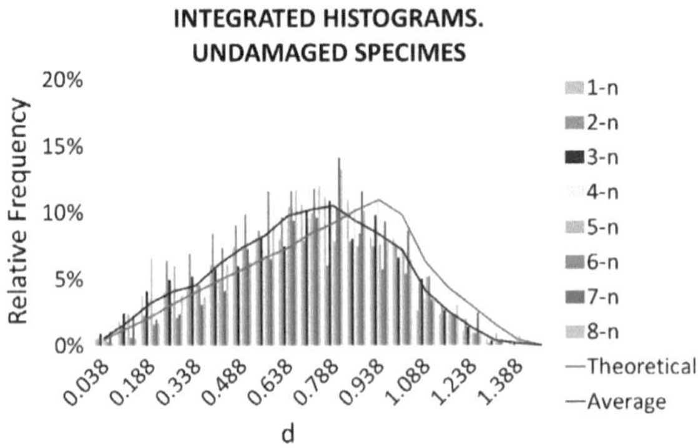

Fig. 9 2D histogram of the intact series specimens

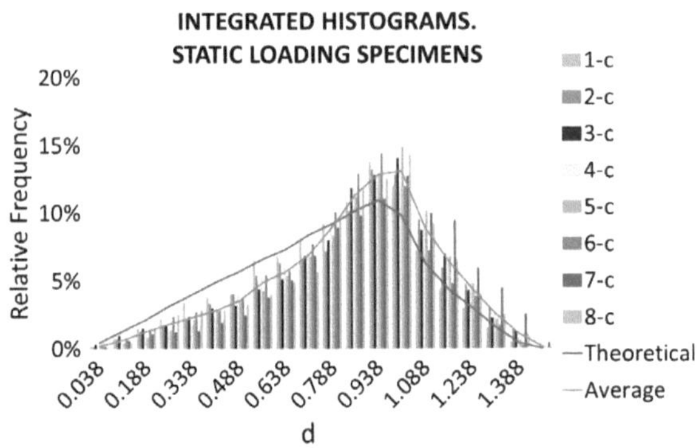

Fig. 10 2D histogram of the monotonic series specimens

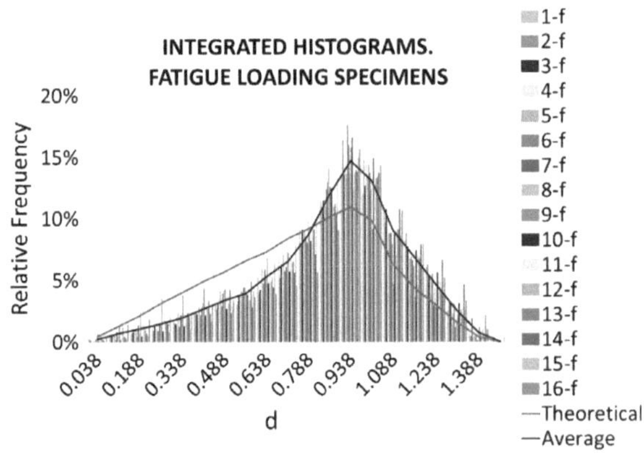

Fig. 11 2D histogram of the cyclic series specimens

Figure 12 shows that theoretical histogram is, in general, inside the confidence interval. This means that, from a statistical point of view, it can be assumed that both distributions are equal, i.e., intact series follow a random distribution. Empty voxel, i.e., pores, are randomly distributed inside the specimens along the d direction.

On the contrary, Figs. 13 and 14 show that theoretical histogram is partially outside the confidence interval. This means that, from a statistical point of view, it can be

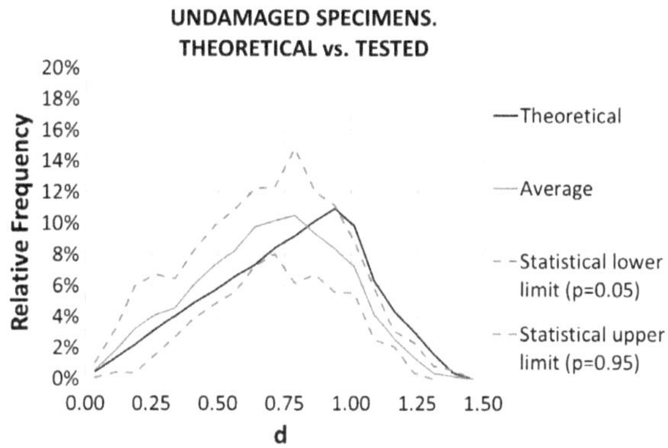

Fig. 12 Comparison between theoretical and intact series

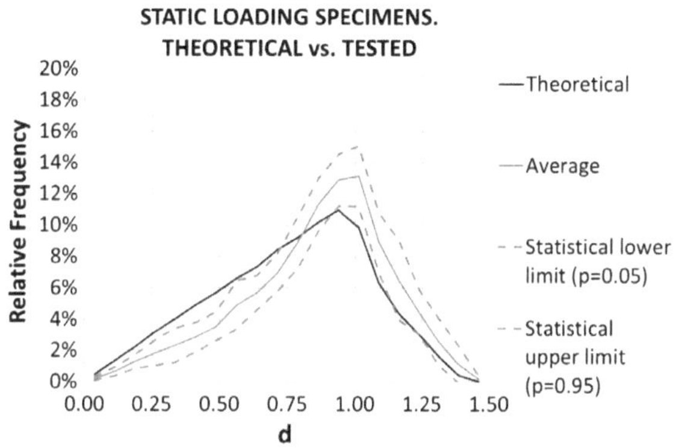

Fig. 13 Comparison between theoretical and monotonic series

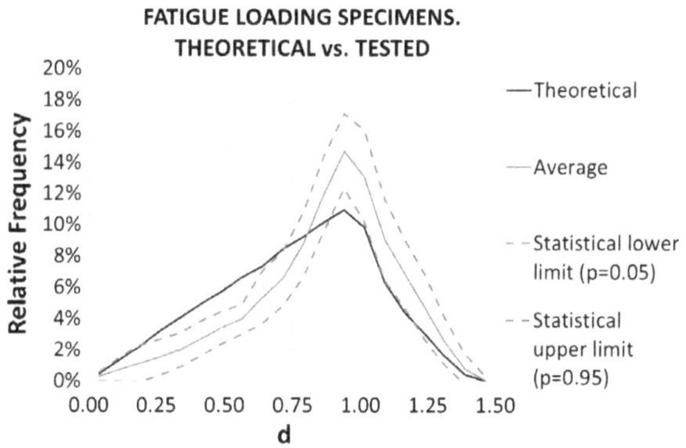

Fig. 14 Comparison between theoretical and cyclic series

assumed that both monotonic and cyclic series do not follow a random distribution along the d direction. In consequence, it can be affirmed that cracks are not randomly distributed inside the specimens.

In case of monotonic and cyclic series, the behavior is quite similar. Between $d = 0$ and $d = 0.8$ approximately, real damage is below theoretical damage. On the

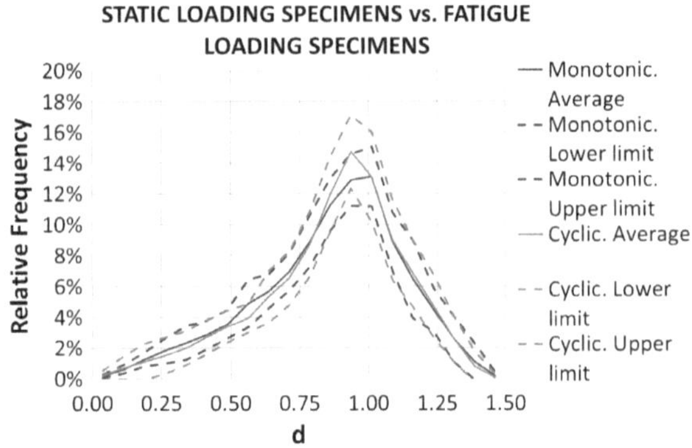

Fig. 15 Comparison between monotonic and cyclic series

contrary, between d = 0.8 and d = $\sqrt{2}$, real damage is above theoretical damage. This means that cracks appear mostly on the edge of the specimens.

Figure 15 show the comparison between the real distributions of monotonic and cyclic series. Additionally, the 90% confidence interval is defined by the upper and the lower limits.

Figure 15 shows that the average monotonic histogram is inside the confidence interval of cyclic series and viceversa. This means that, from a statistical point of view, it can be assumed that both distributions are equal, i.e., the crack pattern of both monotonic and cyclic series are equal along the d direction.

3.3 2D Histograms Along the D Direction

Similarly to the previous case, the 3D histograms were aggregated along the d direction, obtaining 2D histograms which represent the relative frequency of occurrence of empty voxels along the h coordinate, i.e., parallel to the load direction. Figures 16, 17 and 18 show the 2D histograms of the different series. As occurs in the previous case, each histogram show the individual data of the specimens (drawn as bar diagrams) and also the average histogram. Additionally, the histogram belonging to the random distribution is shown, in order to be able to compare real and theoretical distributions.

Figures 19, 20 and 21 show the comparison between the theoretical distribution and the real distributions of intact, monotonic and cyclic series. Additionally, the 90% confidence interval is defined by the upper and the lower limits.

Figures 19, 20 and 21 shows that theoretical histogram is, in general, inside the confidence interval for all series. This means that, from a statistical point of view, it can be assumed that all the distributions are equal, i.e., intact, monotonic and cyclic series follow a random distribution. Empty voxel, i.e., pores, are randomly distributed inside the specimens along the height of the specimens.

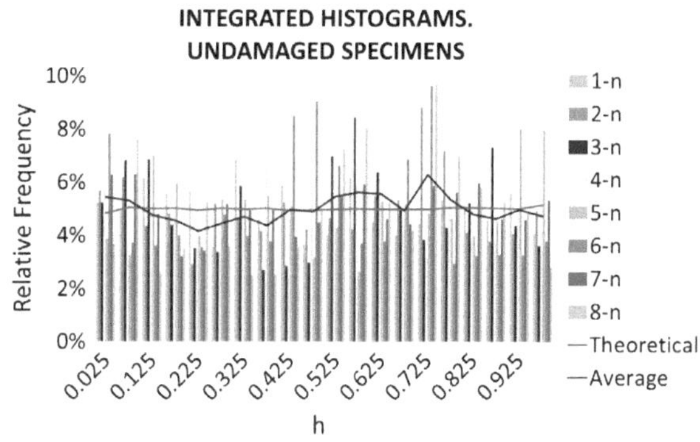

Fig. 16 2D histogram of the intact series specimens

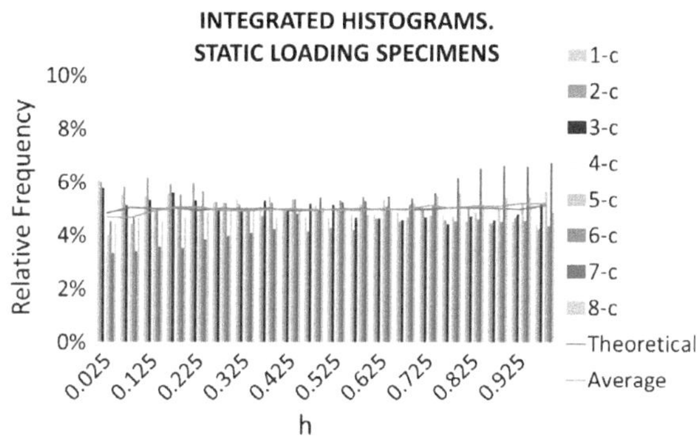

Fig. 17 2D histogram of the monotonic series specimens

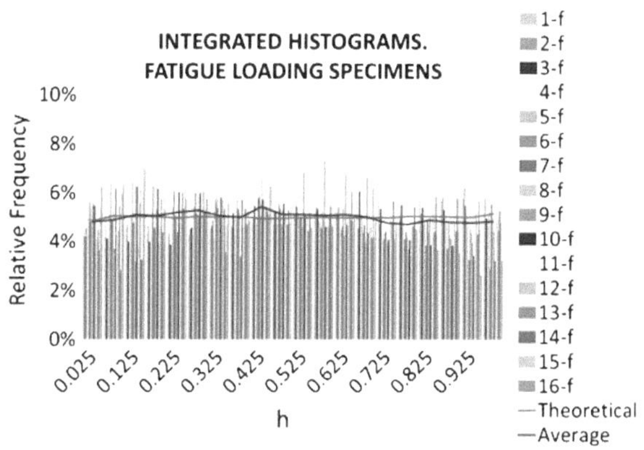

Fig. 18 2D histogram of the cyclic series specimens

Figure 22 shows the comparison between the real distributions of monotonic and cyclic series. Additionally, the 90% confidence interval is defined by the upper and the lower limits.

In concordance with the information shown in Figs. 20, 21 and 22 shows that the average monotonic histogram is inside the confidence interval of cyclic series and viceversa. This means that, from a statistical point of view, it can be assumed

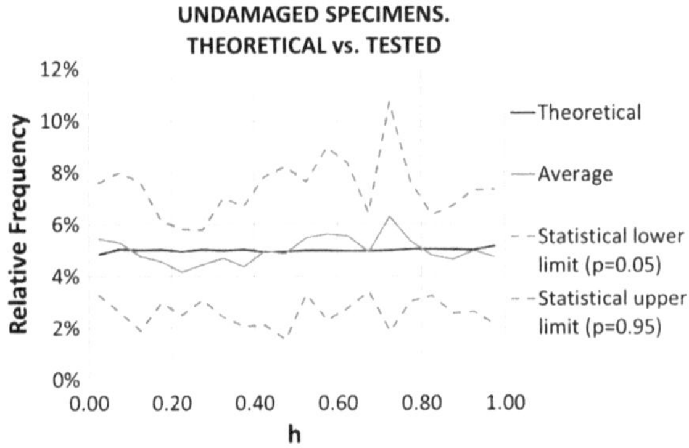

Fig. 19 Comparison between theoretical and intact series

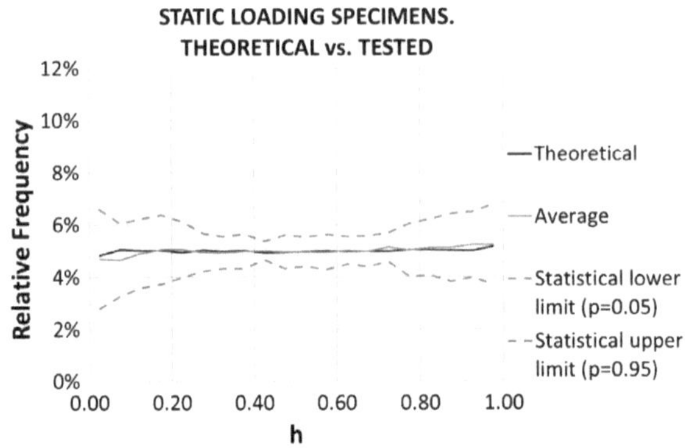

Fig. 20 Comparison between theoretical and monotonic series

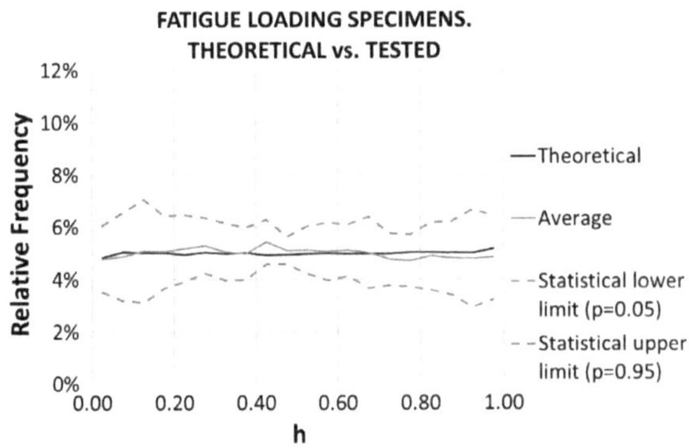

Fig. 21 Comparison between theoretical and cyclic series

that both distributions are equal, i.e., the crack pattern of both monotonic and cyclic series are equal along the h direction.

Considering the information provided by Figs. 14 and 21 it can be affirmed that monotonic and cyclic distributions are equal, which demonstrate the hypothesis shown at the beginning of this paper.

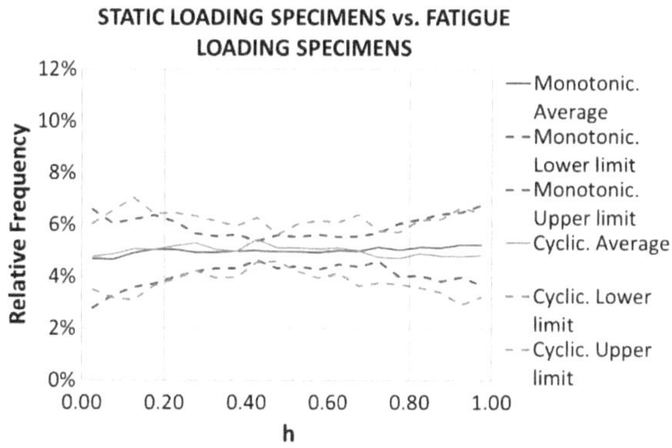

Fig. 22 Comparison between monotonic and cyclic series

4 Summary and Conclusions

The CT-Scan combined with a specific post-processing software is a useful tool to measure the internal damage of concrete specimens. In case of intact specimens, damage refers to pores, while in case of tested specimens, damage refers to pores and cracks.

In this paper, the Circumferential Test is shown. This is a specific protocol, developed by the authors, to measure the damage distribution. Using the data provided by the CT-Scan it is possible to check if damage follows a random distribution or not. In addition, a theoretical random distribution was simulated, using a Monte Carlo model. This simulation was used as the random distribution reference.

Results provided by CT-Scan and later analyzed with the Circumferential Test show that damage (i.e. pores) inside the intact series specimens follow a random distribution, while damage (i.e. pores and cracks) inside the monotonic and cyclic series specimens do not follow a random distribution.

In addition, results show that both monotonic and cyclic follow the same damage distribution. This conclusion demonstrate the hypothesis of convergence to the "initial distribution", which means that the static testing is a particular case of cyclic testing where the fatigue life is equal to 1.

Acknowledgements The authors are grateful for the financial support from the Ministerio de Economía y Competitividad BIA2015-686678-C2-R, Spain, Junta de Comunidades de Castilla – La Mancha, Spain, Fondo Europeo de Desarrollo Regional, gran PEII-2014-016-P and INCRECYT Program.

References

1. Aas-Jackobsen, K.: Fatigue of concrete beams and columns. Ph.D. thesis, University of Trondheim (1970)
2. Bordelon, A.C., Roesler, J.R.: Spatial distribution of synthetic fibers in concrete with x-ray computed tomography. Cement Concr. Compos. **53**, 35–43 (2014). https://doi.org/10.1016/j.cemconcomp.2014.04.007
3. Herrmann, H., Pastorelli, E., Kallonen, A., Suuronen, J.P.: Methods for fibre orientation analysis of x-ray tomography images of steel fibre reinforced concrete (SFRC). J. Mater. Sci. **51**(8), 3772–3783 (2016). https://doi.org/10.1007/s10853-015-9695-4
4. Hsu, T.: J. Am. Concr. Inst. **78**(4), 192–305 (1981)
5. Oesch, T.S., Landis, E.N., Kuchma, D.A.: Conventional concrete and UHPC performance–damage relationships identified using computed tomography. J. Eng. Mech. **142**(12), 04016101 (2016)
6. Pastorelli, E., Herrmann, H.: Virtual reality visualization for short fibre orientation analysis. In: 2014 14th Biennial Baltic Electronic Conference (BEC), pp. 201–204 (2014). https://doi.org/10.1109/BEC.2014.7320591
7. Petkovic, G., Lenschow, R., Stemland, H., Rosseland, S.: Fatigue of high strength concrete. ACI Spec. Publ. **121**(25), 505–525 (1990)
8. Pittino, G., Geier, G., Fritz, L., Hadwiger, M., Rosc, J., Pabel, T.: Computertomografische untersuchung von stahlfaserspritzbeton mit mehrdimensionalen transferfunktionen. Beton- und Stahlbetonbau **106**(6), 364–370 (2011). https://doi.org/10.1002/best.201100009
9. Ponikiewski, T., Katzer, J., Bugdol, M., Rudzki, M.: Steel fibre spacing in self-compacting concrete precast walls by x-ray computed tomography. Mater. Struct. **48**(12), 3863–3874 (2015a). https://doi.org/10.1617/s11527-014-0444-y
10. Ponikiewski, T., Katzer, J., Bugdol, M., Rudzki, M.: X-ray computed tomography harnessed to determine 3D spacing of steel fibres in self compacting concrete (SCC) slabs. Constr. Build. Mater. **74**, 102–108 (2015b). https://doi.org/10.1016/j.conbuildmat.2014.10.024
11. Przybilla, C., Fernández-Cantelli, A., Castillo, E.: Deriving the primary cumulative distributive function of fracture stress for brittle materials from 3- and 4-point bending tests. J. Eur. Ceram. Soc. **31**, 451–460 (2011)
12. Saucedo, L., Yu, R., Medeiros, A., Zhang, X., Ruiz, G.: A probabilistic fatigue model based on the initial distribution to consider frequency effect in plain and fiber reinforced concrete. Int. J. Fatigue **48**, 308–318 (2013)
13. Schnell, J., Schladitz, K., Schuler, F.: Richtungsanalyse von fasern in betonen auf basis der computer-tomographie. Beton- und Stahlbetonbau **105**(2), 72–77 (2010). https://doi.org/10.1002/best.200900055
14. Suuronen, J.P., Kallonen, A., Eik, M., Puttonen, J., Serimaa, R., Herrmann, H.: Analysis of short fibres orientation in steel fibre reinforced concrete (SFRC) using x-ray tomography. J. Mater. Sci. **48**(3), 1358–1367 (2013). https://doi.org/10.1007/s10853-012-6882-4
15. Tepfers, R., Kutti, T.: Fatigue strength of plain, ordinary and lightweight concrete. J. Am. Concr. Inst. **76**(5), 635–652 (1979)
16. Vicente, M., Minguez, J., González, D.: The use of computed tomography to explore the microstructure of materials in civil engineering: from rocks to concrete. In: Halefoglu, D.A.M. (ed.) Computed Tomography-Advanced Applications. InTech (2017). https://doi.org/10.5772/intechopen.69245
17. Vicente, M.A., González, D.C., Mínguez, J.: Determination of dominant fibre orientations in fibre-reinforced high-strength concrete elements based on computed tomography scans. Nondestruct. Test. Eval. **29**(2), 164–182 (2014). https://doi.org/10.1080/10589759.2014.914204
18. Zhang, B., Phillips, D., Wu, K.: Effects of loading frequency and stress reversal on fatigue life of plain concrete. Mag. Concr. Res. **48**(4), 292–305 (1996)
19. Zhao, D., Chang, Q., Yang, J., Song, Y.: A new model for fatigue life distribution of concrete. Key Eng. Mater. **348–349**, 201–204 (2007)

Experimental Investigation on Bending Creep in Cracked UHPFRC

Daniene Casucci, Catherina Thiele and Jürgen Schnell

Abstract Investigations on ordinary fibre-reinforced concrete showed that the time-dependent deformations under tensile load in cracked concrete are larger than the deformations in uncracked concrete. The so-called tensile creep in the cracked cross section depends on some different factors like type of fibres, fibre content, load level, concrete mix, environmental condition, etc. Given the lack of sufficient data about tensile creep in ultra-high performance fibre-reinforced concrete (UHPFRC), a large experimental program financed by the DFG (Deutsche Forschungsgemeinschaft) was started at the University of Kaiserslautern.

1 Introduction

Depending on the fibre content, UHPFRC has often a strain-hardening behaviour. This means that in a tensile or in a bending test, after the formation of the first crack, it is possible to increase the applied load. The concrete full tensile capacity is reached with multiple fine cracks and the material is usually designed in cracked condition. The contribution of the fibres to the tensile strength is so high, that in some cases, it seems reasonable to avoid or reduce the amount of conventional reinforcement. Therefore it is necessary to investigate also the tensile long-term behaviour of the cracked UHPFRC.

D. Casucci (✉) · C. Thiele · J. Schnell
Institute of Concrete Structures and Structural Engineering,
Technische Universität Kaiserslautern, Kaiserslautern, Germany
e-mail: daniele.casucci@bauing.uni-kl.de; Daniele.Casucci@hilti.com

C. Thiele
e-mail: catherina.thiele@bauing.uni-kl.de

J. Schnell
e-mail: juergen.schnell@bauing.uni-kl.de

D. Casucci
Hilti Corporation, Feldkircherstrasse 100, 9494 Schaan, Liechtenstein

H. Herrmann and J. Schnell (eds.), *Short Fibre Reinforced Cementitious Composites and Ceramics*, Advanced Structured Materials 95,
https://doi.org/10.1007/978-3-030-00868-0_2

The aim of this research project, which has been started at the University of Kaiserslautern, is to evaluate the long-term durability and reliability of this material, to estimate the deformations under sustained loads and to find out whether a load limitation has to be imposed in the future design codes. In this paper, besides the adopted test method, some results of the first year of measurements and observation are reported. An overall safe behaviour of steel fibres up to relatively high sustained loads (with respect to the tensile strength in static tests) could be observed.

While the literature concerning tensile creep in ordinary fibre-reinforced concrete has increased in the past years, such literature concerning UHPFRC is still very rare. A large part of the research concerning time-dependent strain in UHPFRC is oriented to the creep in uncracked cross sections and in early age behaviour. This is important to evaluate the internal stress during the hardening phase in restrained conditions for the possibility of cracking [3]. Recent studies concerning the tensile-creep behaviour of high and ultra-high performance concrete may be found in the literature [3, 6–9, 12, 13, 15]. This topic is difficult to investigate since the experiments are complex and the tensile-creep deformations have the same order of magnitude as the shrinkage [7]. The most important parameters of influence, beside the cuing condition [6], are the water-cement ratio [7], the content of silica [3, 6, 7, 15] the age of loading [3, 6, 7, 15] and of course the magnitude of the load [3, 6, 7, 12, 15]. Concerning the magnitude of the tensile creep, is very difficult to make a comparison of the literature, since the testing parameters varies a lot and results vary significantly. While according Rossi [13] tensile creep is smaller than the compressive, it is, according to Kordina of the same size or slightly larger [8]. In Garas [6] it was found that the tensile creep can be even larger than the compressive one. The non-linear behaviour between tensile stress and creep deformation, was found in tension for higher stress/strength ratio than for compression, i.e. of 60–70% [7, 8, 15]. Moreover, it seems that under tensile load the shrinkage has also a different behaviour. Reinhardt et al. [12] discovered that an increase of shrinkage for high performance concrete under tension corresponded with increasing compressive strength.

Concerning the tensile creep in cracked cross sections, an overview of the present literature can be found in [9] or [14]. The tensile creep may be caused by the fibre creep, by a time dependent fibre pullout and by the creep of the cementitious matrix. While polymeric fibres, depending on the material of the polymers, tend to be sensitive to the tensile fibre creep, steel fibres show a more stable behaviour since steel has much lower relaxation. Conventional hooked-end macro-steel fibres engages via a mechanical interlock with the concrete matrix, but show also an increase of the tensile deformation over the time in correspondence with cracks. Nieuwoudt [10] connects this increase to the compressive creep of concrete in the areas of high contact pressure of the fibre hook. Even if a failure is unlikely, the deformations and micro cracking of concrete enable the fibre hook to "slide" and this results in the macroscopic observed time-dependent deformations.

In comparison with the typical hooked-end fibres, UHPFRC steel fibres are usually smaller, with diameter between 0.1 and 0.3 mm, lengths between 6 and 20 mm [2] and are straight and smooth [4]. One of the first investigations of uniaxial tensile creep in UHPFRC specimens with smooth micro-steel fibres was made by Garas [6].

Besides the uncracked specimens used for the tensile creep experiments, he also investigated some pre-cracked specimens. Garas observed a quick stabilization of the tensile creep deformations and failure only for load level above 80% of the static resistance in the first minutes after the reloading in the sustained load test rig. Bărbos [1] found a positive influence of the fibres for the tensile creep of UHPFRC beams with conventional reinforcement. Also in this case, the displacements showed a quick stabilisation. He attributed the creep deformation to the formation of new cracks and not to the widening of the existing ones. Nishiwaki [11] investigated a UHPFRC with a cocktail of short and long hocked-end fibres and observed, beside the formation of new cracks, also a slight widening of the existing ones. No failure occurred and with a creep factor of ca. 0.3 after 28 days, the deformations were very small.

2 Experimental Investigation

Uniaxial and bending tensile sustained load tests with over 60 pre-cracked specimens were started. Some of the specimens were unloaded after a period of about 6 months and tested for the residual strength. Some others are still under load and will remain under observation for the next years. Several parameters have been investigated such as the influence of different load levels, age of loading, the specimen pre-damaging level, fibre volume and fibre length-diameter. Shrinkage, and compressive creep were measured in additional tests in order to identify the contribution of the tensile creep to the bending deformations. This paper will focus only on the results of the bending experiments.

2.1 Experimental Procedure

A standard for these kinds of tests is unfortunately still missing. However, in most of the literature reported in [9] or [14], the method is similar. A schematic representation of the testing procedure is given in Fig. 1. First the specimens are pre-loaded up to a certain load or deformation in a common testing machine (step 1), then are unloaded and installed in a creep test rig (step 2), where they remain under sustained load over a certain time. After that, the sustained load is removed and the specimens are tested up to failure for the determination of the residual strength (step 3). The sustained load is defined as a percentage of the load at the final point of the pre-loading phase.

2.2 Specimens and Test Rigs

The tests were performed in non-notched beams with a four-point bending setup. The four-point bending setup enables the observation of the average deformation of

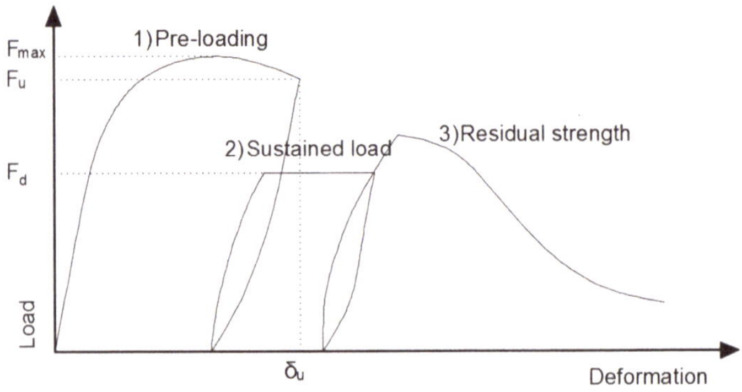

Fig. 1 Schematization of the sustained load creep tests

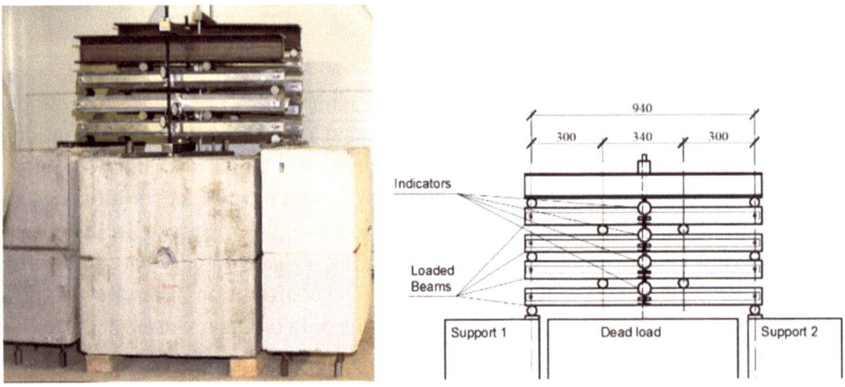

Fig. 2 Sustained load bending tests rig

a larger area of the material compared to notched three-point bending. In the case of a strain-hardening material, it is more advantageous to observe distributed cracking over a large region of the sample than to observe only the small local area around a notch. The beams had a cross-section of 70 × 70 mm, in accordance with the French guideline [4] for the adopted fibre length of 12.5 mm. The span was increased to a length of 940 mm and the central part with constant bending moment was 340 mm long. Besides enabling the observation of a large portion of the material, the larger length was also useful in reducing the required dead load. The vertical displacement was measured with two precision indicators, fixed on both sides in the mid-span of the beams. For the four-point bending tests, eight separate stacks containing three to four specimens each were constructed, as represented in Fig. 2.

Table 1 Mix design of the investigated UHPFRC

Material	Content (kg/m^3)
Cement CEM 52.5 N	728
Water	80
Sand DM 0.125 / 0.5	816
Quartz flour	510
Microsilica suspension	226
Super plasticizer	29.7
Steel fibre (2% or 4%)	164 / 328

2.3 *Concrete Mix*

The adopted concrete mix is indicated in the Table 1. The steel fibres are manufactured by the company Krampeharex® and are straight brass alloy coated steel wires. Brass alloy coated fibres are the most common and guarantee a good connection with the concrete paste [10]. The water-cement ratio, under consideration of the water in the silica suspension, is 0.265 and the overall water-binder ratio is 0.23. A heat treatment for 48 h at 90 °C was performed two days after casting of the specimens. The concrete had a compressive strength of 150–160 MPa if cured in laboratory conditions and 180–190 MPa after the heat treatment.

2.4 *Test Program*

A "reference" combination of 12 beams was tested extensively with scheduled sustained loads 40, 60, 80 and 90% of the residual strength in the pre-loading tests (series 1). The bending creep in non-cracked concrete was measured with two beams loaded at 40 and 60% of the residual strength in the pre-loading tests (series 2). Unfortunately, with loads higher than 60% of the residual strength in the pre-loading tests, cracks would appear making difficult the evaluation of the results.

The influence of fibre slenderness was investigated in four beams containing fibres with 0.400 mm diameter. These fibres have the same length of 12.5 mm as the fibres used in the other experiments, but had a length-diameter ratio of 31.3 instead of 71.4 of the 0.175 mm diameter fibres (series 3).

For investigating the effect of the fibre content, four tests with 4% fibre volume and sustained load levels of 40 and 80% were performed (series 4).

The heat treatment minimizes the effects of creep and shrinkage. However, a test series on specimens without heat treatment was also performed. A total of four specimens were cured in laboratory conditions, with humidity of about 50% and other three were cured in water and sealed with wet cloth (series 5 and 6). This measure was intended to avoid any kind of drying shrinkage.

Table 2 Sustained load test program

Series	Investigated parameter	Fibre length; Diameter; content	Sustained load [%] (scheduled load)	Number of tests
1.	Load level	12.5; 0.175; 2%	40 / 60 / 80 /90	5 / 3 / 2 / 2
2.	Bending creep in uncracked concrete	12.5; 0.175; 2%	(Ca. 40 / 60) without pre loading	1 / 1
3.	Length / diameter ratio	12.5; 0.40; 2%	40 / 80	2 / 2
4.	Fibre content	12.5; 0.175; 4%	60	2
5.	Without heat treatment	12.5; 0.175; 2%	40 / 80	3 / 1
6.	Water curing and sealing	12.5; 0.175; 2%	40% / without pre loading	2 / 1
7.	Concrete age in cracked and non-cracked concrete	12.5; 0.175; 2%	50% / without pre loading	4 / 2
	Total			33

Additionally, three specimens were also tested at ages of 2 and other three at the ages of 13 days. Two of these specimens were pre-loaded and one not. Also in this case the load was only 50% of the residual strength, not to induce cracks in the uncracked specimens (series 7).

Six dog-bone shaped specimens and six beams were tested for each different concrete mix according to the French guideline [2]. With these tests, a deformation level for the pre-loading phase was defined. This deformation was a vertical displacement of 1.5 mm for the beams with 0.40 mm diameter fibres and of 4 mm for the beams with 0.175 mm fibres. These values were chosen so that the UHPFRC would have a well-developed crack pattern during the experiments and so that at the same time the residual strength would still be considerable, larger than 70% of the maximum load.

Since the specimens were piled, more specimens received the same external load. Therefore the actual load in test differed from the scheduled indicated in Table 2.

The test program for the sustained load tests on beams is summarized in Table 2. Here the fibre length, diameter, fibre content and the load level are given. A total of 33 bending creep tests were performed.

2.5 Compressive Creep and Shrinkage

To estimate the contribution of the compressive creep to the deformation of the beam specimens, 27 cylinders with a diameter of 104 mm were tested for a period between two weeks and four months. The tests were performed with three load levels of 0.25,

Table 3 Naming of the specimens

No.		Significance
1.	Specimen number	PK01, PK02, ...
2.	Curing	Wb.: heat treatment curing W.: water curing k. B.: curing in laboratory condition
3.	Additional parameters	n. Vb.: without pre load Wb.: 2T: age at load beginning of two days 13T: age at load beginning of thirteen days sealed: sealed specimen 0.4Ø: fibre diameter of 0.4 mm instead of 0.175 mm 4%: fibre content of 4% instead of 2%
4.	Load level	L = XX%: load level

0.45 and 0.65 f_c with a concrete age of 2, 13 and 28 days. For each combination, a sealed specimen, an unsealed specimen and a heat-treated specimen were tested.

3 Test Results

In this section, some representative results of the creep tests will be shown in order to draw the first conclusion of this experimental test program. The following Table 3 illustrate the naming adopted in the following paragraphs for the single specimens.

3.1 Bending Tests

Figure 3 shows the creep deformations of all the specimens of test series 1. Despite the large scatter, a slight correlation between the load level and the bending creep deformation could be observed. One of the specimens (PK05Wb. in Fig. 3) collapsed after 23 days with a load of 79% of the residual strength at the end of the pre-loading phase. During the pre-loading, this specimen exhibited a dramatic decrease in resistance after having reached the maximum load. An examination of the cross section after the collapse showed a particularly unfavourable fibre-orientation characteristics. Fibres were indeed laying almost parallel to the crack plane without offering an effective crack bridging (Fig. 4). Other specimens contained within the same stack as specimen PK05Wb (for instance PK07Wb.) were also affected by this collapse.

Figure 5 shows a comparison between cracked and non-cracked specimens. In this diagram, only specimens that can be directly compared for charge and load levels are shown. The pre-cracked UHPFRC showed always larger creep deflection than the uncracked ones.

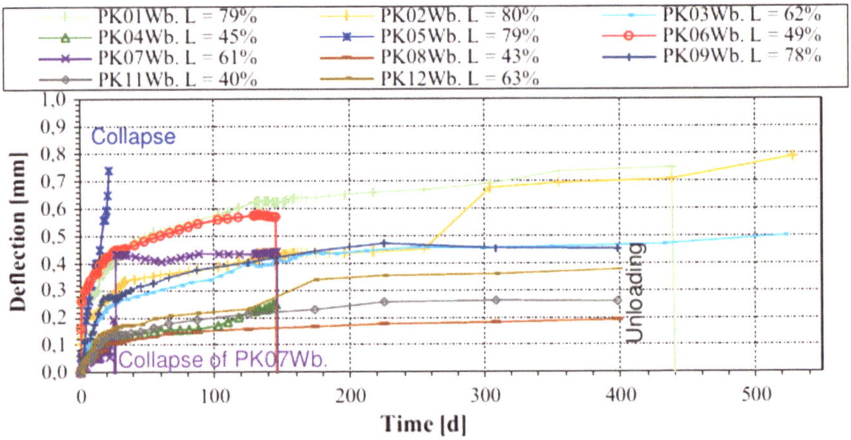

Fig. 3 Creep deflection for specimens in test series 1 of Table 2

Fig. 4 Cross section of the collapsed specimen PK05Wb

Figure 6 shows the results of the specimens with age of 2 and 13 days. These results indicate that increasing deformation is present within samples of lesser age as compared to samples of greater age. A clear difference between the deformation levels of pre-cracked and non-cracked samples was again visible within this dataset, however the difference between non-cracked and pre-cracked concrete remains of the same proportion. This suggest that deformations in cracked concrete are mainly due to the creep of the compression area.

The sealing of the specimens seems not to have a pronounced influence on the beams as shown in Fig. 7 for two different load level of about 40 and 60%. Specimens with 0.4 mm fibre diameter showed slightly smaller creep deflections than those with 0.175 mm (Fig. 8).

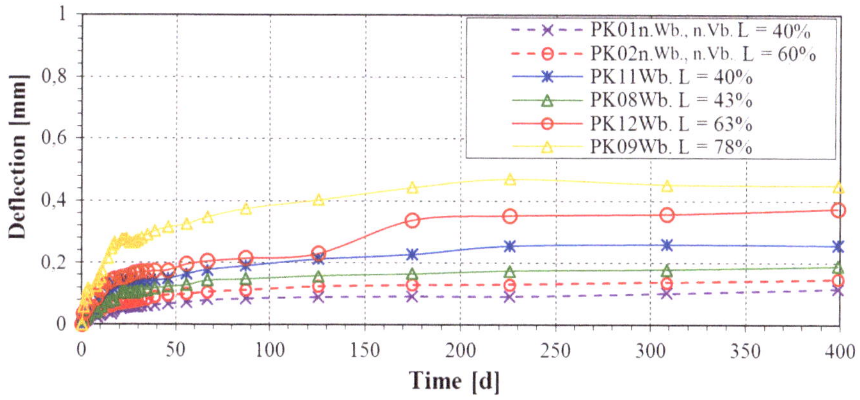

Fig. 5 Creep deflection for cracked and non-cracked specimens (series 2 of Table 2)

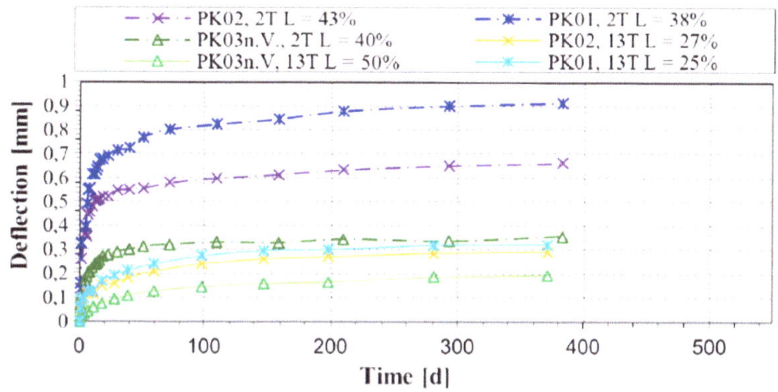

Fig. 6 Creep deflection for specimens of series 2 of Table 2 (these specimens were cured in laboratory conditions)

In general, a stabilization of the displacements occurred within the first 30 days. Some of the specimens were unloaded after 147 days and residual strength tests were performed. Although the largest displacements occurred within the first month, a continuous increase could be stated for the whole observation period.

3.2 *Residual Strength Tests*

Currently, only a few number of specimens were unloaded and tested for the residual strength. These were the specimens with the lower sustained load levels. Indeed in the creep deformation was more stable and a collapse on long term was unlikely. The load displacement curve in the residual strength tests continued the envelope of the pre-load curve. This indicated a good state of the material and the capacity

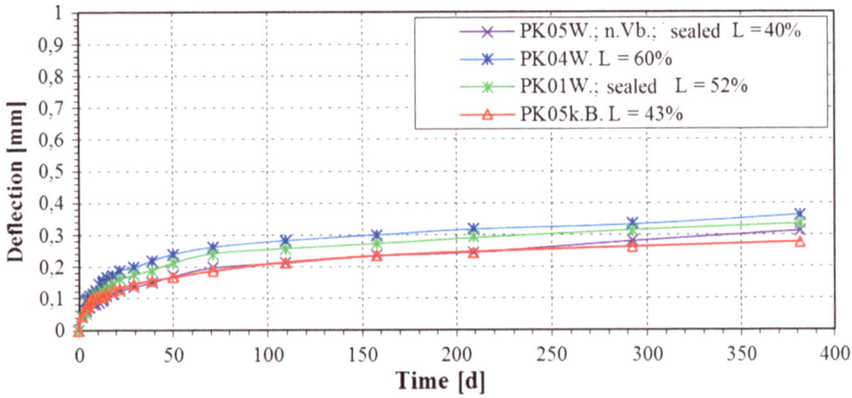

Fig. 7 Creep deflection for specimens of series 6 in Table 2

Fig. 8 Creep deflection for specimens with 0.4 mm fibre diameter (series 3 in Table 2)

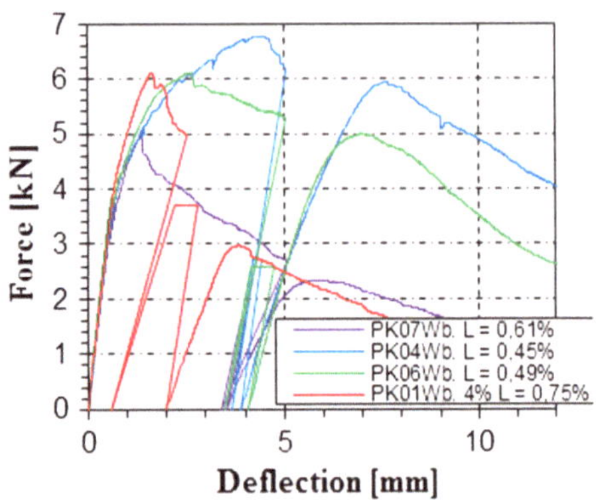

to sustain the designed loads also after the sustained load. Figure 9 shows some of the load displacement curves including the pre-load, sustained load and residual strength tests. The only cases in which the residual strength load displacement curve was lower than expected observing the pre-loading curve, were those where the pre-loading was ending in a steeply descending branch. An example of that is the PK01Wb.Ø0.4 or PK01Wb. 4% of Fig. 9.

3.3 *Compressive Creep and Shrinkage*

Figure 10 reports the basic creep coefficient $\varphi_b(t,t_0)$ for some of sealed specimens. The $\varphi_b(t,t_0)$ is defined according the following equation:

$$\varepsilon_{cc}(t, t_0) = \sigma(t_0)[\varphi_b(t, t_0)]/E_c \tag{1}$$

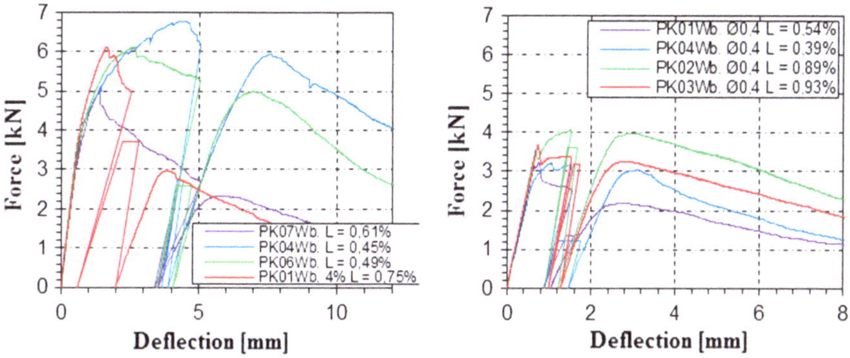

Fig. 9 Residual strength tests on heat-threated beams (naming of the specimens defined in Figs. 3, 5 and 6)

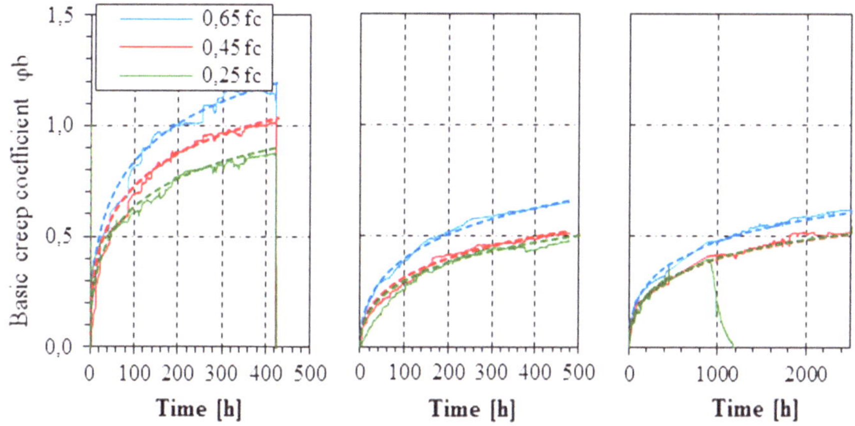

(a) 2 Days (b) 8 Days (c) 28 Days

Fig. 10 Compressive creep, at the age of 2, 8 and 28 days. The dashed lines indicate the progression according to the experimental method of EN 1992-1-2

whereby $\varepsilon_{cc}(t,t_0)$ is the strain, $\sigma(t_0)$ is the compressive stress, E_c elasticity modulus of concrete, and t_0 the age of the concrete at the beginning of the loading. The results were extrapolated with the equation of the experimental method of EN 1992-1-2 [5]. The dashed lines in Fig. 10 indicates this approximation. In Fig. 10 one can observe that the basic creep reduces significantly as the age of the sample at the time of loading increases from 2 to 28 days. The results indicate that the stress-deformation non-linearity limit rises with the age of the concrete for values below 0.65 f_c. This can be observed in Fig. 10 from the slight difference between the specimen loaded with 0.25 and 0.45 f_c and the larger difference between these lower loads and that of 0.65 f_c. Figure 11 shows the test rig for the compressive creep and the basic and drying shrink age.

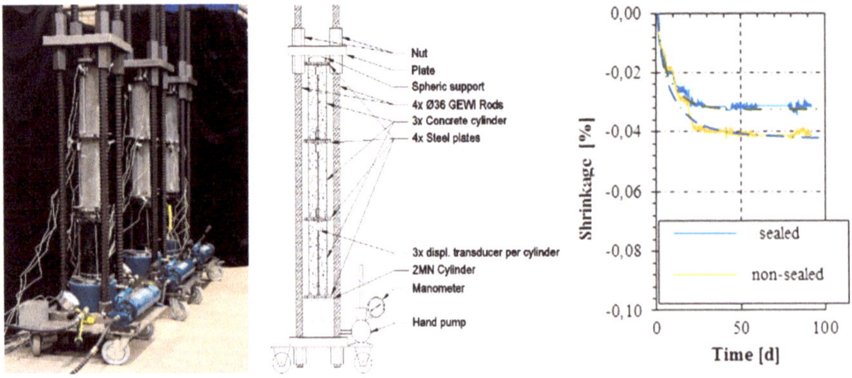

Fig. 11 Compressive creep test rig (left and centre) and deformation diagram (right) of the Shrinkage in sealed specimens (blue) and non-sealed specimens (yellow). The dashed lines indicate the profile according the EN 1992-1-2

4 Conclusions

For an investigation of the creep behaviour of concrete, a few months are not usually enough to gain an exhaustive knowledge of the material. However, it seems that UHPFRC exhibits a stabilization of the time dependent deformation in a relatively short time, even in cracked condition. That means that the working principle of the fibres seems not to be subject to degradation over the time. The only collapsed specimen was a one with an unfavourable fibre orientation and a high sustained load of 79% of the residual strength after pre-loading. This indicates that the fibre orientation has also a relevance for the creep deformation and ultimate resistance. This parameter, although difficult to investigate, should somehow be taken into account. One possible way is correlating it to the post crack behaviour of the specimens.

The effect of the load level was found to be low. At this time, no further conclusions about the effects of loading can be drawn given the scatter within the data of the cracked specimens.

The sustained load had almost no impact on the residual strengths. More important seems to be the post-cracking behaviour and how fast the load displacement curve drops at the unloading point. In the bending tests, no influence of sealing was observed. Tests of samples containing higher fibre content and different fibre diameters did not exhibit any significantly different load performance than the standard reference samples.

Acknowledgements This research was possible thanks to the Deutsche Forschungsgemeinschaft (DFG), which financed and implemented a new research-training group (RTG) named "Stochastic Models for Innovations in the Engineering Sciences" at the University of Kaiserslautern.

References

1. Bărbos, G.A.: Long-term behavior of ultra-high performance concrete (UHPC) bended beams. Procedia Technol. **22**, 203–210 (2016)
2. Bétons fibrés à ultra-hautes performances and Association Française de Gènie Civil: Afgc richtlinie. Documents scientifique et technique (2013)
3. Bissonnette, B., Pigeon, M.: Tensile creep at early ages of ordinary, silica fume and fiber reinforced concretes. Cement Concr. Res. **25**(5), 1075–1085 (1995)
4. Deutscher Ausschuss für Stahlbeton: Dafstb heft 561 ultrahochfester beton—sachstandsbericht (2008)
5. Deutsches Institut für Normung e. V.: Din en 1992-2 eurocode 2: Eurocode 2: Design of concrete structures-Part 1-2: General rules-structural fire design and german version en 1992, 1–2 (2004)
6. Garas, V.Y.: Multi-scale investigation of the tensile creep of ultra-high performance concrete for bridge applications. Ph.D., Georgia Institute of Technology (2009)
7. Kamen, A., Denarié, E., Sadouki, H., Brühwiler, E.: Uhpfrc tensile creep at early age. Mater. Struct. **42**(1), 113–122 (2009)
8. Kordina, K.: Beton unter langzeit-zugbeanspruchung. Bautechnik **76**(6), 479–488 (1999)
9. Kusterle, W.: Creep of fibre-reinforced concrete—flexural test on beams. In: Proceedings of Fibre Concrete (2015)
10. Nieuwoudt, P.D.: Time-dependent behaviour of cracked steel fibre-reinforced concrete: from single fibre level to macroscopic level. Ph.D., Stellenbosch (2016)
11. Nishiwaki, T., Kwon, S., Otaki, H., Igarashi, G., Shaikh, F.U., Fantilli A.P.: Experimental study on time-dependent behavior of cracked UHP-FRCC under sustained loads. In: Serna, P., Llano-Torre, A. (eds.) Creep Behaviour in Cracked Sections of Fibre-reinforced Concrete. Proceedings of the International RILEM Workshop FRC-CREEP (2016)
12. Reinhardt, H.W., Rinder, T.: Tensile creep of high-strength concrete. J. Adv. Concr. Technol. **4**(2), 277–283 (2006)
13. Rossi, P., Tailhan, J.L., Le Maou, F., Gaillet, L., Martin, E.: Basic creep behavior of concretes investigation of the physical mechanisms by using acoustic emission. Cement Concr. Res. **42**(1), 61–73 (2012)
14. Serna, P., Llano-Torre, A., Cavalaro, S.H.P.: v.14 creep behaviour in cracked sections of fibre-reinforced concrete. In: Proceedings of the International RILEM Workshop FRC-CREEP 2016. Dordrecht (2016)
15. Switek, A., Denarié, E., Brühwiler, E.: Modeling of viscoelastic properties of ultra high performance fiber reinforced concrete (UHPFRC) under low to high tensile stresses. In: ConMod 2010: Symposium on Concrete Modelling (2010)

An Initial Report on the Effect of the Fiber Orientation on the Fracture Behavior of Steel Fiber Reinforced Self-Compacting Concrete

Heiko Herrmann, Andres Braunbrück, Tanel Tuisk, Oksana Goidyk and Hendrik Naar

Abstract This paper presents a report about work in progress of research on the influence of the fiber orientations on the tensile strength of steel fiber concrete. Different fiber orientations in different parts of a structural element are caused by the casting process. Here, as an example, a small plate was cast of self-compacting concrete containing hooked-end steel fibers. The plate was cut into three beams, which in turn have been subjected to X-ray Computed Tomography scanning to obtain fiber orientations and to three-point bending test, to assess the tensile strength and fracture behaviour.

1 Introduction

During the past decades fiber concrete has become a material construction companies are more and more interested in, due to its potential to form a ductile concrete. This, in turn, has also created interest among researchers from different universities across the world to better understand the material. Due to the short fibers being introduced to the concrete mass at the mixing stage, the otherwise (mostly) homogeneous and isotropic concrete becomes inhomogeneous and anisotropic depending on the spatial and orientational distribution of the fibers [1–3]. For construction companies these effects are unpleasant companions in the "field use" of the material. Inhomo-

H. Herrmann (✉) · A. Braunbrück · O. Goidyk
Department of Cybernetics, Tallinn University of Technology, Tallinn, Estonia
e-mail: hh@cens.ioc.ee

A. Braunbrück · T. Tuisk · H. Naar
Department of Civil Engineering and Architecture, Tallinn University of Technology, Tallinn, Estonia
e-mail: andres@cens.ioc.ee

H. Naar
e-mail: hendrik@mec.ee

T. Tuisk
e-mail: tanel.tuisk@ttu.ee

H. Herrmann and J. Schnell (eds.), *Short Fibre Reinforced Cementitious Composites and Ceramics*, Advanced Structured Materials 95,
https://doi.org/10.1007/978-3-030-00868-0_3

geneity and anisotropy make the material much more challenging to understand and introduce an additional level of complexity into the theoretical description.

In this paper, we present an ongoing study, similar to [4, 5], with the target of comparing the experimental behaviour of fiber concrete under bending load with different theoretical frameworks. Here, only the initial setup of the study and initial outcomes are presented, the comparison with theoretical models is future work.

1.1 Theoretical Work

So far, several theoretical models have been presented by researchers involving different levels of information about the fiber orientations [2, 3, 6–11]. In these, the level of fiber orientation information taken into account varies greatly, from only fiber amount, over average projected length (orientation number) [6], orientation profile [3] to orientation or alignment tensors and derived orientation parameters [2, 8, 9, 11, 12].

Some countries have developed (draft) guidelines for fiber concrete use in construction, among these Denmark [13], Germany [14], Sweden [15], USA [16, 17], Russia [18, 19] and Austria [20]; where the Austrian guideline is valid only against old national building code, and effectively defunct since Eurocode-2 was adopted. An international approach is [21].

The Swedish standard, for example, only allows the use of fiber concrete in load bearing structures with additional reinforcement. In general, the production method, i.e. the casting method, and its influence on fiber orientations is not taken into account in the standards. This shows, that there is still need for more research.

1.2 Overview of Testing Methods

Several non-destructive methods have been proposed that allow the estimation of the micro-structure, such as the fiber orientation and distribution inside the concrete matrix or damage. These methods include x-ray computed tomography scan (CT) [22–27], image analysis [28], conductometric (AC-IS) [29], electromagnetic [30, 31], ultrasonic [32] and acoustic analysis. However, these methods are not easily implemented in real life [22, 28, 33]. After successful application of non-destructive methods in medical research, they have been increasingly used in the investigation of cementitious composites due to their accuracy, efficiency and non-destructive nature [24].

Currently, assessment of the fiber distribution, its orientation inside the matrix and influence on the mechanical properties of concrete have been accepted as fundamental challenges because of the time- and resource-consuming processes [34, 35]. The majority of obstacles are connected with inability to observe the fiber orientation and distribution within concrete mass immediately after casting.

The appearance and widespread use of steel fiber concrete have aroused interest to investigate the material by new non-destructive testing (NDT) methods for inner microstructure analysis and evaluation of the bonds between aggregates, cement paste and steel fibers. NDT techniques are used widely for a long time, but specifically to be applied in civil engineering not too many are available [36]. These methods have been used during more than three decades to evaluate the condition of a structure and to acquire precise information about the concrete samples. Their main advantages are higher reliability, effectiveness and avoidance of material damage during testing.

The majority of researchers have pointed out that it is essential to apply different NDT methods to fiber cementitious materials. These methods vary from very simple to technically complicated depending on the purpose, including CT, conductometric (AC-IS), electromagnetic, ultrasonic, and acoustic emission methods. This section reviews the main NDT methods applied on concrete structures.

Propagation and reflection of various rays, such as CT scan, through a structure, may be also used to study the amount of damage experienced by concrete structures [22, 23, 28, 37]. An in-depth analysis by Liu et al. [24] using micro-CT technique proves that the influence of size and quantity of aggregate on fiber orientation in cementitious materials is more significant than the influence on spacial distribution of fibers. According to Ponikiewski et al. [22], there is an exponential distribution of the fiber angles with respect to the beam main axis. Moreover, CT scan allows the automatic air pore segmentation, provides data on the position and shape properties and evaluation of orientation of steel fibers. Promentilla et al. [23] demonstrate the use of CT technique in the microstructure studies of cement-based materials in relation to the durability performance of these materials.

Another method for detection of fiber density and orientation is the Alternating Current Impedance Spectroscopy (AC-IS) that make use of the electromagnetic properties of steel fibers with a probe designed for this purpose. Ferrara et al. [29] report that the local average concentration and orientation of the steel fibers can be assessed by measuring the variation of the probe inductance. According to his study, the method is easy to implement: a probe is brought in contact with the specimen, without any particular care about the quality of the contact, nevertheless the test results are repeatable with low uncertainties.

Among other electromagnetic testing methods for steel fiber concrete is surface electromagnetic sensor technique [38]. This method requires access to one surface and uses a radio wave system that is able to measure steel fiber distribution, concentration and orientation in the concrete [38].

In Electrical Resistance Tomography (ERT) several electrodes are connected to the specimen surface, that both send alternating currents to the specimen and also record the resulting voltages from the specimen. The measured data allows to recreate this conductivity distribution of the whole specimen volume [30]. Steel reinforcement bars or metallic fibers, cracks, air voids, changes in moisture, distribution of chloride, all influence conductivity [30].

The main idea of the inductive method proposed by Torrents et al. [31] is to measure the variation in self-induction while the specimen is turned with respect to three orthogonal axes. As a rough estimate, the average of the three measurements is considered independent of the fiber orientation [31].

With regard to concrete structures, ultrasound has been used for the detection of cracks and corrosion and also for thickness measurements [36]. Schickert et al. [32] have reported about Ultrasonic Synthetic Aperture Focusing Technique (SAFT). From many pulse-echo measurements a tomographic image is constructed [32]. For concrete with distributed damage a nonlinear ultrasonic technique is developed [39]. Ultrasound attenuation and the appearance of the second harmonic are used to characterise distributed damage [39]. These techniques can be used not only to investigate the microstructure of the material but with recent developments like direct-sequence spread-spectrum ultrasonic evaluation (DSSSUE) changes in properties after damage of the material like shape, density, homogeneity and acoustic velocity may be detected in large and complicated structures [36].

The working principle of acoustic emission (AE) is that a crack is detected the moment it appears, because a propagating crack generates high frequency sound waves and if these waves are extracted from the background noise, the condition of the structure is monitored [36]. AE is very sensitive technique and is able to pick up microscopic changes in the material [36]. During monitoring, the AE sensors of AE work continuously but passively. Unlike other NDE techniques —like radiography, ultrasound, eddy current—AE does not need much preparation time to be set up and is therefore cost-efficient.

In addition, Grosse et al. [40] have pointed out that AE gives us the complete timeline of the damage process without any additional manipulations to the specimen, AE is able to determine the fracture type, size, energy and crack orientation.

The presented various NDT methods are based on different theories, principles and applications, with their advantages and disadvantages. NDT methods have been shown to play a significant role in the assessment and investigation of the microstructure of building materials. Generally, the main advantage of the NDT techniques is the ability to examine the matrix of cementitious materials and reveal hidden defects and acquire data about fiber orientation and distribution without causing damage to the specimen. To achieve better assessment and precise results, it is required to combine several methods. Based on experience, theory and purposes, the best combination of the NDT techniques can be selected for diagnostics of concrete structures.

2 Materials and Methods

To investigate the influence of fiber orientations on the fracture behaviour of fiber concrete, three beam specimen have been prepared by casting a small plate of self-compacting fiber concrete and cutting it into beams. The beams have been subjected to x-ray computed tomography scanning to obtain information about fiber placement and orientation, and to three point bending tests to obtain the stress-strain diagrams. The methods are described in detail below.

Table 1 Data of used fibers: Semtu WireFib 80/60; amount used 25kg/m^3

Length	60 mm
Diameter	0.75 mm
Aspect ratio	80
Number of fibers/kg n.	4600
Tensile strength	>1000 MPa
Coating	Uncoated
Steel quality	EN 10016-2 C9

Table 2 Concrete data

Casting day		30.09.16
Slump (diameter) of concrete		750 mm
Temperature		20 °C
Density and Compressive strength at 3 day	2470 kg/m^3	48.2 MPa
Density and Compressive strength at 7 day	2460 kg/m^3	61.1 MPa
Density and Compressive strength at 28 day	2470 kg/m^3 2470 kg/m^3	71.2 MPa 67.5 MPa

2.1 *Fresh Concrete*

A self-compacting concrete mix was prepared according to a recipe by the producer. Steel fibers of 60 mm length and 0.75 mm diameter with hooked ends (see Table 1) have been added at the mixing stage and mixed with the concrete according to the data sheet.

Cubes for compression tests have been prepared according to the usual procedure, the test results are presented in Table 2.

2.2 *Casting of a Small Plate*

To test the influence of the fiber orientation on the mechanical properties of the fiber concrete, especially on the peak load and post-cracking behaviour, a small plate has been cast. The size of the plate was $(L \times W \times H)$ 90 cm × 60 cm × 20 cm.

The plate was cast from one end with a 10 cm wide inlet and flow directed along the length of the plate. This casting was chosen to enable comparison with the fluid simulations and "simulation experiments" (see [41, 42]). As mentioned before, the concrete was self-leveling concrete and the slump-test had a diameter of 75 cm.

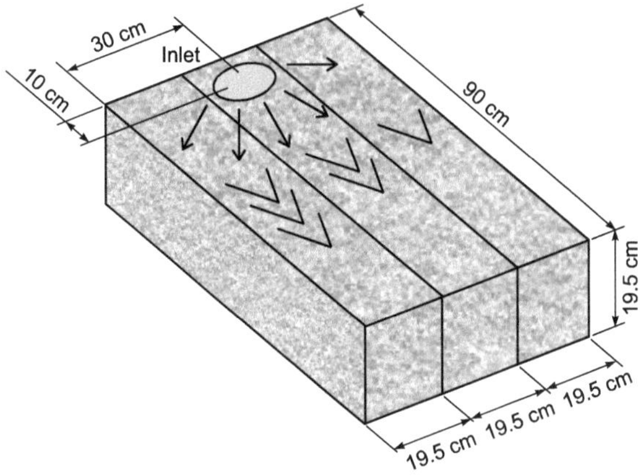

Fig. 1 Cutting beams from the plate

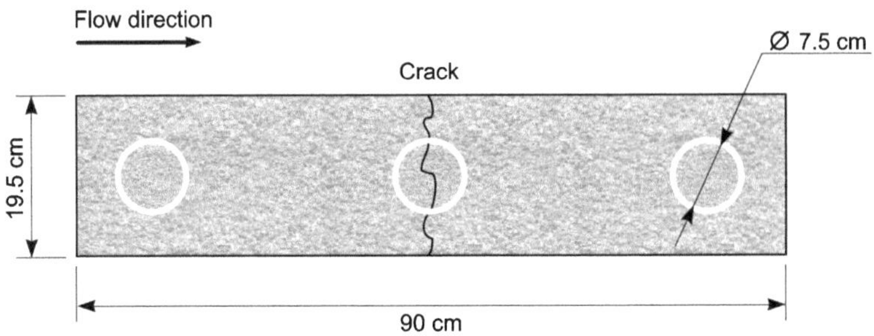

Fig. 2 Position of cylinders

2.3 Preparing Beams from the Plate

This plate has been cut into three beams of size $(L \times W \times H)$ 90 cm × 19.5 cm × 19.5 cm (see Fig. 1). According to the computer simulations and "simulation experiments" (see [41, 42]) the outer beams should have the fibers mostly oriented along the main axis of the beam, while in the middle beam the fibers should be mostly perpendicular to the main axis.

2.4 Drilling of Cylinders

After the mechanical testing of the beams, cylinder samples were drilled out of the beams (see Fig. 2), to enable a better tomography scanning of some areas, especially of the crack area, and also to perform split-tests (see Sect. 4.2).

3 Fiber Orientations Measurement by X-Ray Computed Tomography

X-ray computed tomography has become an efficient and reliable method to obtain information about the inner structure of materials. It can be used on the macro and down to the nano scale, depending on the sample size and x-ray attenuation of the scanned objects. Previous studies have used x-ray computed tomography to obtain information about aggregates, porosity [43, 44] and fiber positions and orientations [25–27, 42, 45–48].

In this study, fiber orientations in a small plate are of interest. The plate was cut into beams and later, after bending tests, cylinders have been drilled from the beams. The beams and cylinders have been subjected to x-ray computed tomography scanning using medical scanners. The devices used and results are discussed in the following.

3.1 X-Ray CT of the Beams

The beams have been scanned using a GE Medical Systems "Discovery STE", the reconstruction of the volume was performed with 0.6 mm resolution, Table 3 summarizes the parameters. The size of the beams (cross section 19.5 cm × 19.5 cm) turned out to be slightly above the attenuation limit for this scanner, fibers can only be clearly identified close to the surfaces of the beam, not in the inner region. The fibers in the bottom layer of the beams can be seen in Fig. 3. Visually the fiber orientations coincide with the expectations from a casting into a slippery formwork [4, 41, 42].

From the orientation ellipses shown in Fig. 4d, f one can see, that the orientation is more favourable to take tension stresses in the outer regions of the plate, namely the left side of beam 3 and the right side of beam 1, while in the center region of the plate (beam 2) the fiber orientation distribution is more isotropic with a tendency to alignment perpendicular to the tension stress.

The numbers of fibers per beam segment and per cell in the beams differ from 11 to 29, with the highest numbers in the central beam (beam 2), see Fig. 4g, h, i.

3.2 X-Ray CT of the Cylinders

Due to the problems with separating fibers from the matrix in the CT scans of the beams, cylinders have been drilled from the beams after the bending tests, to identify fiber orientations in the bulk and in the crack plane. The scans were again performed using a medical CT scanner, but this time a GE Medical Systems "Discovery CT750 HD" was used. The parameters of the scan are summarized in Table 4.

Table 3 Scanning and reconstruction parameters used for beams

Manufacturer	GE MEDICAL SYSTEMS
Manufacturers model name	Discovery STE
Full fidelity	CT_LIGHTSPEED
Acquisition group length	412
Scan options	HELICAL MODE
Slice thickness	0.625000
Data collection diameter	500.000000
Reconstruction diameter	317.000000
X Ray tube current	400
Filter type	BODY FILTER
Convolution kernel	BONEPLUS
Single collimation width	0.625
Total collimation width	40
Table feed per rotation	20.625
Spiral pitch factor	0.515625
Rows	512
Columns	512
Pixel spacing	0.619141/0.619141
Bits allocated	16
Bits stored	16
High bit	15
Pixel representation	Signed

An example of the data received from the CT scan is given in Fig. 5. The volume image is thresholded to show only the greylevel corresponding to fibers, but as one can see, some aggregate particles have almost the same attenuation, and therefore greylevel, as the fibers.

4 Strength Experiments and Results

Two types of mechanical tests have been performed: three-point bending tests of the beam specimen and splitting tests on cylinders drilled from the beams after the bending tests.

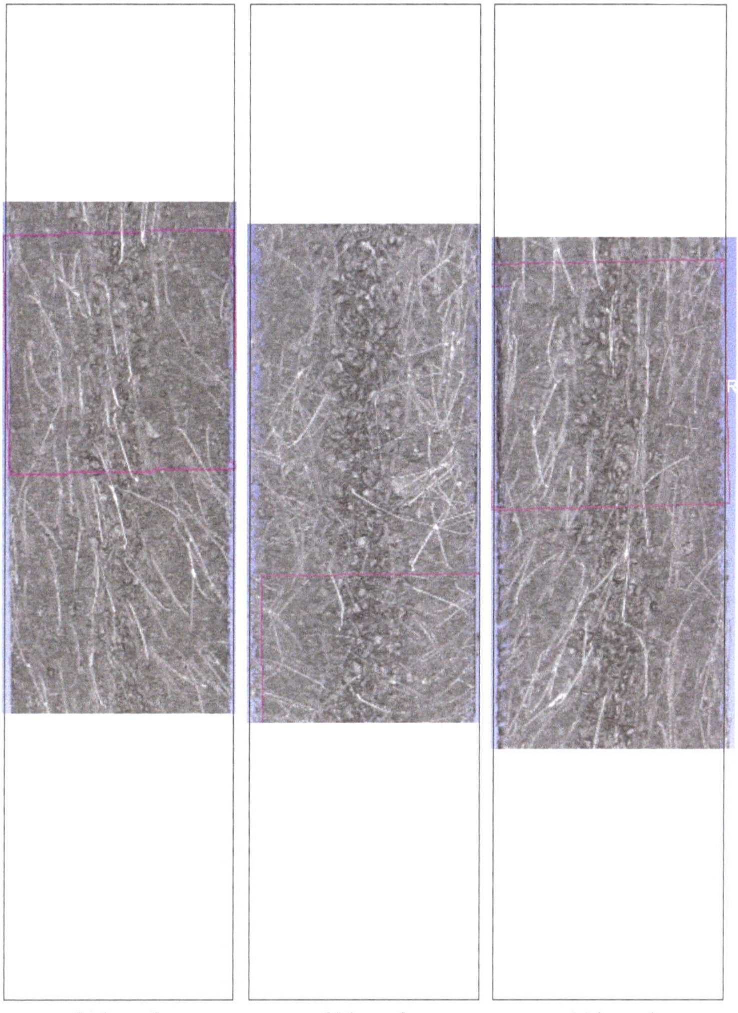

Fig. 3 Bottom layer of the central section of the beams in x-ray CT, the volume image is thresholded to show the fibers in the bottom layer

4.1 Bending Test

The three-point bending tests of the fiber-concrete beam specimens was performed on *Zwick Roell z250* strength testing machine. Three beams with dimensions $90 \times 19.5 \times 19.5$ cm were tested. The two vertical supports A and B were $L = 0.78$ m apart, each being positioned 6 cm inward from the end of the beam, Fig. 6. The line-load F was applied at the mid cross-section C (Fig. 7).

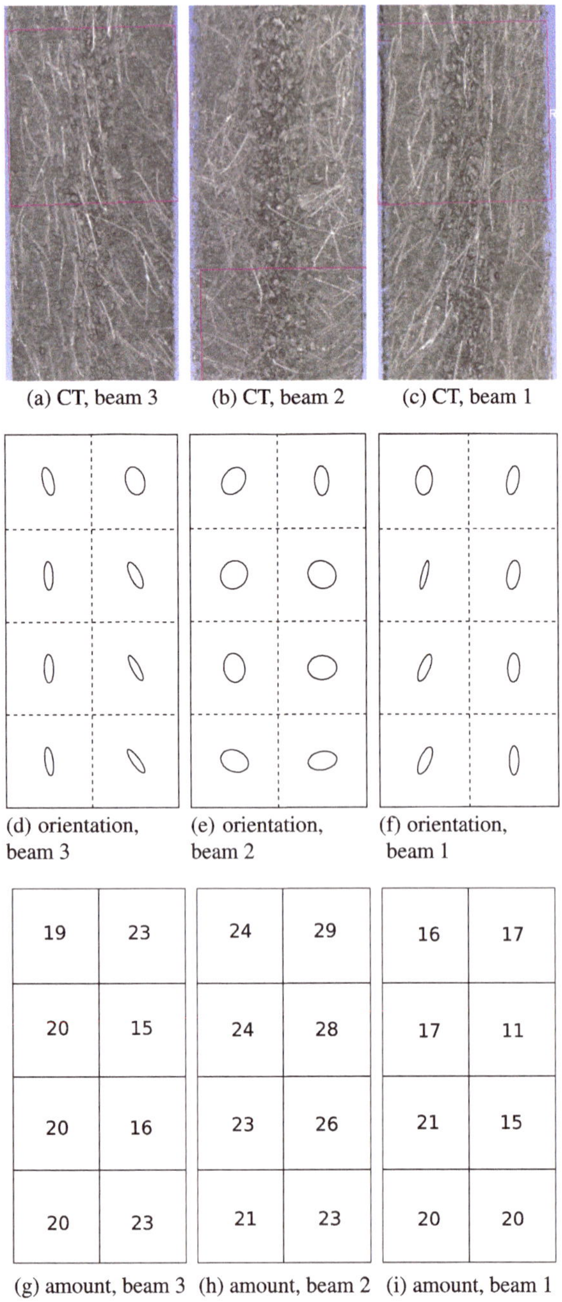

Fig. 4 CT image of center parts of beams thresholded to show fibers in bottom layer, fiber orientation ellipses and fiber amount per cell in the three beams; beam 1: 137 fibers; beam 2: 198; beam 3: 156

Table 4 Scanning and reconstruction parameters used for cylinders

Manufacturer	GE MEDICAL SYSTEMS
Rows	512
Columns	512
Bits allocated	16
Bits stored	16
High bit	15
Pixel representation	Signed
X Ray tube current in micro amps	300000.00
Acquisition type	SPIRAL
Single collimation width	0.62
Total collimation width	40
Spiral pitch factor	0.52

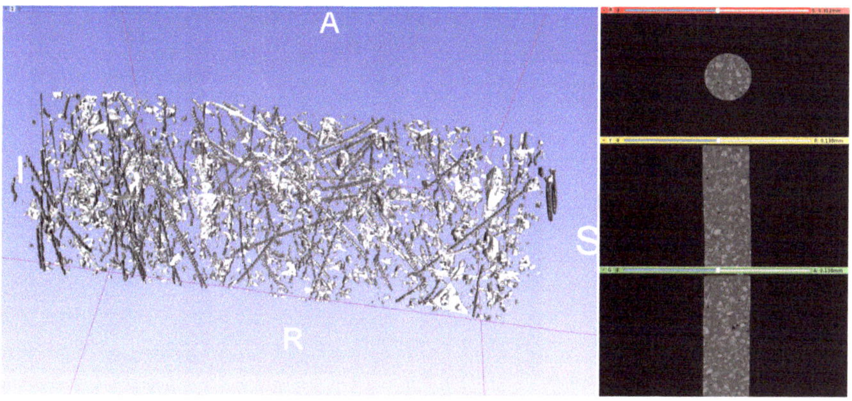

Fig. 5 Volume and slice images of a CT scanned cylinder

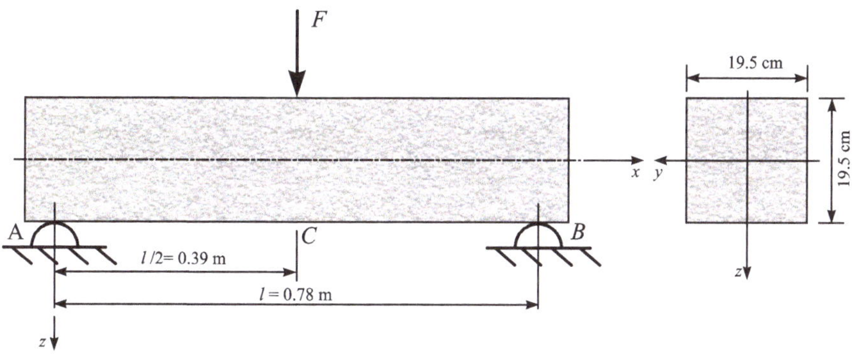

Fig. 6 Bending test setup

Fig. 7 Photo of the bending test showing a beam sample in the testing machine with the frame for displacement measurements and displacement sensors

The recorded force-displacement curves are presented in Fig. 8 where the maximum force value before cracking is $F_1 = 45.36$ kN, $F_2 = 44.06$ kN and $F_3 = 43.18$ kN for beams 1, 2, and 3, respectively. The formula to calculate the maximum bending stresses in a cross-section is $\sigma^{M_y} = \frac{M_y}{W_y}$ that we apply in the cross-section C where the bending moment is given by $M_y = FL/4$ and the section modulus is given by $W_y = (wh^2)/6$.

Flexural fracture strength (in MPa):

$$\sigma^{M_y} = \frac{M_y}{W_y} = \frac{FL}{4}\frac{6}{w * h^2} \tag{1}$$

$$\sigma_{\text{fs}} = \frac{1.5 * F * L}{w * h^2} \tag{2}$$

with F maximum force at cracking (in N) , L length between supports (in mm), w the width of the sample (in mm) and h the height of the sample (in mm).

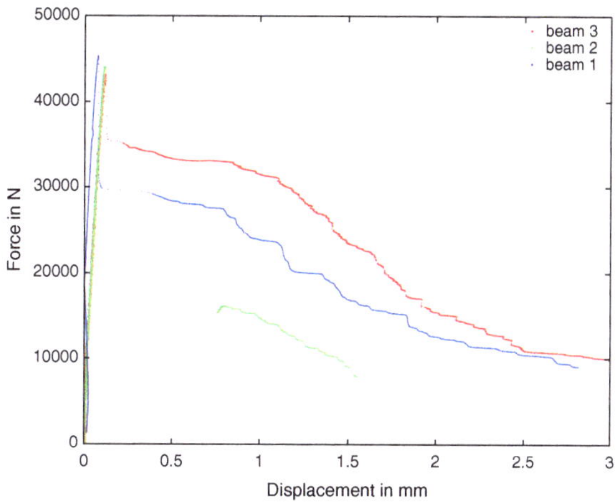

Fig. 8 Force displacement curves for the three-point bending test of the three beams

Table 5 Flexural fracture strength of the three beams; the average was $\overline{\sigma}^{M_y} = 6.97$ with a standard-deviation of 0.18

Beam	max. force in N	FFS in MPa
1	45364.7	7.16
2	44058.4	6.95
3	43178.6	6.81

For a cross-section of 19.5 × 19.5 cm the maximum stress is $\sigma_1^{M_y} = 7.16$ MPa, $\sigma_2^{M_y} = 6.95$ MPa, $\sigma_3^{M_y} = 6.81$ MPa for beams 1, 2, and 3, respectively, see Table 5. Over the three experiments we get the averaged maximum stress $\overline{\sigma}^{M_y} = 6.97$ MPa.

Comparing the results of the bending test with the number and orientation of fibers in the immediate bottom layer, one can note the following: (1) the peak-strength shows no obvious correlation with the fiber amount in/near the crack-region, (2) though beam 2 has the highest number of fibers in the cells next to the crack, it shows the weakest post-cracking, (3) though having the highest number of fibers, their orientations is the least beneficial in beam 2, this seems to show correlation with the post-cracking behaviour. Taking both fiber amount and orientation into account one can note: beam 1 and 3 have both similar orientation distributions and beam 3 has a higher number of fibers, beam 3 also shows the best post-cracking behaviour of the three beams.

4.2 Split Test

The split test was performed with a *Matest IT Tech Cyber Plus Evlolution* testing machine on a single specimen that was drilled out from the intact part of the beam

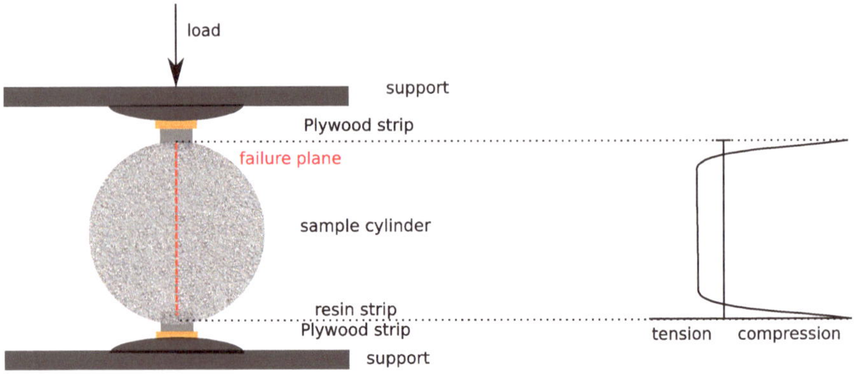

Fig. 9 Schematic of split test

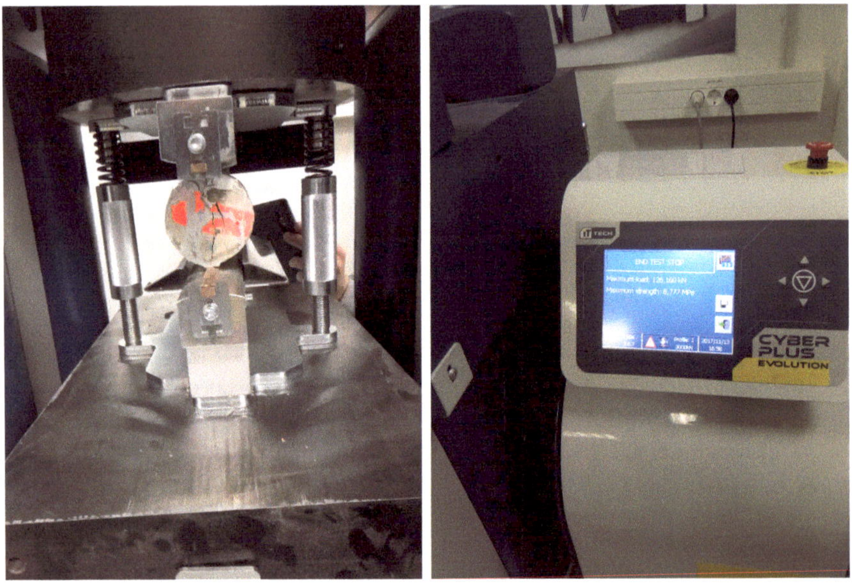

Fig. 10 Split test

sample after the three-point bending test. The cylinder specimen had the average diameter $d = 74$ mm and height $l = 195$ mm. In addition to the regular setup of a concrete split test a special epoxy glue interlayer was prepared on two opposite sides of the cylinder to facilitate a smoother distribution of loads over the cylinder surface, Fig. 9.

The maximum force during the split test was $P = 126160$ N, Fig. 10. The splitting tensile strength σ in MPa may be calculated as [49]:

$$\sigma = \frac{2P}{\pi l d} = \frac{2 \cdot 126160}{\pi \cdot 74 \cdot 195} \frac{N}{mm^2} = 5.57 \text{ MPa} \tag{3}$$

with P the maximum applied load in N, l the length im mm and d the diameter in mm.

For our sample a splitting tensile strength $\sigma = 5.57$ MPa resulted.

5 Future Work

The casting of a larger plate, from which more beams with similar to identical fiber orientation distributions can be cut, is underway. Also, for this casting at the same time more reference samples for compression tests of pure concrete and fiber concrete will be cast, additionally a reference plate out of pure concrete will be cast to produce reference beams for the bending tests. This will improve statistics and provide error margins.

6 Conclusion

In this preliminary study the contribution of fiber orientation on the strength of steel fiber reinforced concrete was investigated. Three-point bending tests have been performed on beam samples and special attention has been paid to the fiber amount and orientation distribution in the bottom layer. While the results give no conclusive evidence that the peak strength depends on the amount or orientation of fibers, the post-cracking behaviour shows dependence on fiber orientations and amount. The following has been noted: the beam with the highest amount of fibers in the bottom layer (highest tensile stress) showed the weakest post-cracking behaviour, it had also the least beneficial fiber orientation distribution. Two beams with similar beneficial fiber orientation distributions showed better post-cracking behaviour, the one with a slightly higher fiber count was the stronger one among the two in the post-cracking regime.

Acknowledgements The authors gratefully acknowledge the funding by the Estonian Research Council by the exploratory research grant PUT1146.

We also thank Maria Kremsreiter who helped during her Erasmus+ internship at the Institute of Cybernetics. Therefore: With the support of the Erasmus+ programme of the European Union. Thanks to E-Betoonelement, especially Aare Lessuk, Rasmus-R. Marjapuu and Sergei Graf, for preparing the experiment plate.

References

1. Eik, M., Lõhmus, K., Tigasson, M., Listak, M., Puttonen, J., Herrmann, H.: DC-conductivity testing combined with photometry for measuring fibre orientations in SFRC. J. Mater. Sci. **48**(10), 3745–3759 (2013)
2. Herrmann, H.: An improved constitutive model for short fibre reinforced cementitious composites (SFRC) based on the orientation tensor. In: Generalized Continua as Models for Classical and Advanced Materials, pp. 213–227. Springer International Publishing, Cham (2016)
3. Laranjeira, F., Grünewald, S., Walraven, J., Blom, K., Molins, C., Aguadoa, A.: Characterization of the orientation profile in fiber reinforced concrete. Mater. Struct. **44**(6), 1093–1111 (2011)
4. Žirgulis, G., Švec, O., Sarmiento, E.V., Geiker, M.R., Cwirzen, A., Kanstad, T.: Importance of quantification of steel fibre orientation for residual flexural tensile strength in FRC. Mater. Struct. 1–17 (2015)
5. Švec, O., Žirgulis, G., Bolander, J.E., Stang, H.: Influence of formwork surface on the orientation of steel fibres within self-compacting concrete and on the mechanical properties of cast structural elements. Cement Concr. Compos. **50**, 60–72 (2014)
6. Grünewald, S.: Performance-based design of self-compacting fibre reinforced concrete. Ph.D. thesis, Technische Universiteit Delft (2004)
7. Eik, M., Puttonen, J., Herrmann, H.: An orthotropic material model for steel fibre reinforced concrete based on the orientation distribution of fibres. Compos. Struct. **121**, 324–336 (2015)
8. Eik, M., Herrmann, H., Puttonen, J.: Orthotropic constitutive model for steel fibre reinforced concrete: linear-elastic state and bases for the failure. In: Kouhia, R., Mäkinen, J., Pajunen, S., Saksala, T. (eds.) Proceedings of the XII Finnish Mechanics Days: 4–5 June 2015, Tampere, Finland, pp. 255–260 (2015)
9. Herrmann, H., Eik, M., Berg, V., Puttonen, J.: Phenomenological and numerical modelling of short fibre reinforced cementitious composites. Meccanica **49**(8), 1985–2000 (2014)
10. Krasnikovs, A., Kononova, O., Khabbaz, A., Machanovsky, E., Machanovsky, A.: Post-cracking behaviour of high strength fiber concrete prediction and validation. World Acad. Sci. Eng. Technol. **59**, 988–992 (2011)
11. Eik, M., Puttonen, J., Herrmann, H.: The effect of approximation accuracy of the orientation distribution function on the elastic properties of short fibre reinforced composites. Compos. Struct. **148**, 12–18 (2016)
12. Herrmann, H., Beddig, M.: Tensor series expansion of a spherical function for use in constitutive theory of materials containing orientable particles. Proc. Est. Acad. Sci. **67**(1), 73–92 (2018) (Open-Access CC-BY-NC 4.0)
13. Kasper, T., Tvede-Jensen, B., Stang, H., Mjoernell, P., Slot, H., Vit, G., Thrane, L.N., Reimer, L.: Design guideline for structural applications of steel fibre reinforced concrete. Technical report, SFRC Consortium (2014)
14. Deutscher Ausschuss für Stahlbeton: DAfStb-Richtlinie. Beuth Verlag GmbH, Berlin (2010)
15. SIS.: Fibre concrete-design of fibre concrete structures (Swedish standard – SS 812310:2014) (2014)
16. ACI Committee 544.: 544.4r-88 design considerations for steel fiber reinforced concrete. Technical report, American Concrete Institute (2002)
17. ACI Committee 544.: 544.5r-10: Report on the physical properties and durability of fiber-reinforced concrete. Technical report, American Concrete Institute (2010)
18. SNIP.: 52-104-2006–steel fibre reinforced concrete structures design. Technical report, SNIP, Moskau (2007)
19. SNIP.: 52-104-2009–steel fibre reineorced concrete structures design. Technical report, SNIP, Moskau (2010)
20. Österreichische Bautechnikvereinigung. Richtlinie faserbeton, 2008. Österreichische Vereinigung für Beton- und Bautechnik
21. Fib.: Model Code 2010. International Federation for Structural Concrete (fib) (2012)

22. Ponikiewski, T., Gołaszewski, J., Rudzki, M., Bugdol, M.: Determination of steel fibres distribution in self-compacting concrete beams using x-ray computed tomography. Arch. Civil Mech. Eng. **15**, 558–568 (2015)
23. Promentilla, M.A.B., Sugiyama, T., Shimura, K.: Threedimensional imaging of cement-based materials with x-ray tomographic microscopy: visualization and quantification. In: International Conference. Microstructure Relat Durab Cem Compos, vol. 61, pp. 1357–1366 (2008)
24. Liu, J., Li, C., Liu, J., Cui, G., Yang, Z.: Study on 3D spatial distribution of steel fibers in fiber reinforced cementitious composites through micro-ct technique. Constr. Build. Mater. **48**, 656–661 (2013)
25. Pastorelli, E., Herrmann, H.: Time-efficient automated analysis for fibre orientations in steel fibre reinforced concrete. Proc. Est. Acad. Sci. **65**(1), 28–36 (2016)
26. Herrmann, H., Pastorelli, E., Kallonen, A., Suuronen, J.-P.: Methods for fibre orientation analysis of x-ray tomography images of steel fibre reinforced concrete (SFRC). J. Mater. Sci. **51**(8), 3772–3783 (2016)
27. Suuronen, J.-P., Kallonen, A., Eik, M., Puttonen, J., Serimaa, R., Herrmann, H.: Analysis of short fibres orientation in steel fibre reinforced concrete (SFRC) using x-ray tomography. J. Mater. Sci. **48**(3), 1358–1367 (2013)
28. Liu, J., Sun, W., Miao, C., Liu, J., Li, C.: Assessment of fiber distribution in steel fiber mortar using image analysis. J. Wuhan Univ. Technol. Mater. Sci. Ed. **27**, 166–171 (2012)
29. Ferrara, L., Faifer, M., Toscani, S.: A magnetic method for non destructive monitoring of fiber dispersion and orientation in steel fiber reinforced cementitious composites—Part 1: method calibration. Mater. Struct. 1–15 (2011)
30. Karhunen, K., Seppänen, A., Lehikoinen, A., Monteiro, P.J.M., Kaipio, J.P.: Electrical resistance tomography imaging of concrete. Cement Concr. Res. **40**, 137–145 (2010)
31. Torrents, J.M., Blanco, A., Pujadas, P., Aguado, A., Juan-García, P., Sánchez-Moragues, M.Á.: Inductive method for assessing the amount and orientation of steel fibers in concrete. Mater. Struct. **45**(10), 1577–1592 (2012)
32. Schickert, M.: Progress in ultrasonic imaging of concrete. Mater. Struct. **38**, 807–815 (2005)
33. Grigaliunas, P., Kringelis, T.: SCC flow induced steel fiber distribution and orientation, non-destructive inductive method. In: 11th European Conference on Non-Destructive Testing. Prague, Czech Republic (2014)
34. Zhou, B., Uchida, Y.: Fiber orientation in ultra high performance fiber reinforced concrete and its visualization. In: 8th International Conference on Fracture Mechanics of Concrete and Concrete Structures (2013)
35. Svec, O., Skocek, J., Olesen, J.F., Stang, H.: Fibre reinforced self-compacting concrete flow simulations in comparison with l-box experiments using carbopol. In: 8th Rilem International Symposium on Fibre Reinforced Concrete (2012)
36. Rens, K.L., Wipf, T.J., Klaiber, F.W.: Review of nondestructive evaluation techniques of civil infrastructure. J. Perform. Constr. Facil. **11**, 152–160 (1997)
37. Akhtar, S: Review of nondestructive testing methods for condition monitoring of concrete structures. J. Constr. Eng. (2013)
38. Al-Mattarneh, H.: Electromagnetic quality control of steel fiber concrete. Constr. Build. Mater. **73**, 350–356 (2014)
39. Shah, A.A., Hirose, S.: Nonlinear ultrasonic investigation of concrete damaged under uniaxial compression step loading. J. Mater. Civil Eng. **11**, 476–484 (2009)
40. Grosse, C.U., Reinhardt, H.W., Finck, F.: Signal-based acoustic emission techniques in civil engineering. J. Mater. Civil Eng. **15**, 274–279 (2003)
41. Herrmann, H., Lees, A.: On the influence of the rheological boundary conditions on the fibre orientations in the production of steel fibre reinforced concrete elements. Proc. Est. Acad. Sci. **65**(4), 408–413 (2016) (Open-Access CC-BY-NC 4.0)
42. Herrmann, H., Goidyk, O., Braunbrück, A.: Influence of the flow of self-compacting steel fiber reinforced concrete on the fiber orientations, a report on work in progress. In: Short Fiber Reinforced Cementitious Composites and Ceramics. Springer (2018)

43. Ponikiewski, T., Katzer, J., Bugdol, M., Rudzki, M.: Determination of 3D porosity in steel fibre reinforced SCC beams using x-ray computed tomography. Constr. Build. Mater. **68**, 333–340 (2014)
44. du Plessis, A., le Roux, S.G., Guelpa, A.: Comparison of medical and industrial x-ray computed tomography for non-destructive testing. Case Stud. Nondestruct. Test. Eval. **6**, 17–25 (2016)
45. Vicente, M.A., González, D.C., Mínguez, J.: Determination of dominant fibre orientations in fibre-reinforced high-strength concrete elements based on computed tomography scans. Nondestruct. Test. Eval. **29**(2), 164–182 (2014)
46. Ponikiewski, T., Katzer, J., Bugdol, M., Rudzki, M.: Steel fibre spacing in self-compacting concrete precast walls by x-ray computed tomography. Mater. Struct. **48**(12), 3863–3874 (2015)
47. Ponikiewski, T., Katzer, J., Bugdol, M., Rudzki, M.: X-ray computed tomography harnessed to determine 3D spacing of steel fibres in self compacting concrete (SCC) slabs. Constr. Build. Mater. **74**, 102–108 (2015)
48. Minguez, M.A.V.J., Gonzalez, D.C.: Image data processing to obtain fibre orientation in fibre-reinforced elements using computed tomography scan. In Short Fiber Reinforced Cementitious Composites and Ceramics. Springer (2018)
49. ASTM International.: Standard test method for splitting tensile strength of cylindrical concrete specimens. Technical report, ASTM, 2004. Designation: C 496/C 496M – 04

Non-destructive Evaluation of the Contribution of Polymer-Fibre Orientation and Distribution Characteristics to Concrete Performance during Fire

Tyler Oesch, Ludwig Stelzner and Frank Weise

Abstract Although concrete itself is not a combustible material, concrete mixtures with high density, such has high-performance concretes (HPCs), are susceptible to significant damage during fires due to explosive spalling. Past research has shown that the inclusion of polymer fibres in high density concrete can significantly mitigate this fire damage. The exact mechanisms causing this increased spalling resistance are not yet fully understood, but it is thought that the fibres facilitate moisture transport during fire exposure, which in turn contributes to relief of internal stresses in the spalling-susceptible region. In this study, X-ray Computed Tomography (CT) was applied to observe the interaction between polymer fibres and cracking during thermal exposure. For this purpose, two concrete samples containing different polymer fibre types were subjected to incremental application of a defined thermal exposure. CT images were acquired before and after each thermal exposure and powerful image processing tools were used to segment the various material components. This enabled a detailed analysis of crack formation and propagation as well as the visualization and quantification of polymer fibre characteristics within the concrete. The results demonstrated that the orientation of both fibres and cracks in polymer-fibre reinforced concrete tend to be anisotropic. The results also indicated that crack geometry characteristics may be correlated with fibre orientation, with cracks tending to run parallel to fibre beds. Clear quantitative relationships were also observed between heating and increasing cracking levels, expressed in terms of both crack surface area and crack volume.

T. Oesch (✉) · L. Stelzner · F. Weise
Bundesanstalt für Materialforschung und –prüfung,
Federal Institute for Materials Research and Testing, 12205 Berlin, Germany
e-mail: tyler.oesch@bam.de

L. Stelzner
e-mail: ludwig.stelzner@bam.de

F. Weise
e-mail: frank.weise@bam.de

H. Herrmann and J. Schnell (eds.), *Short Fibre Reinforced Cementitious Composites and Ceramics*, Advanced Structured Materials 95,
https://doi.org/10.1007/978-3-030-00868-0_4

1 Introduction

1.1 Performance of Concrete During Fire

Although concrete itself is not a combustible material, concrete structural components are susceptible to explosive spalling during fire. The investigation and control of this phenomenon is very important since the spalling of individual components can have a significant effect on the overall fire resistance of structures. Structural fire resistance, in turn, has major implications both to the safety of first responders during fires as well as the costs associated with structural repair following a fire.

It has been demonstrated that this spalling behaviour is at least partially related to the presence of moisture within the concrete material [9, 28]. Spalling behaviour is also increased for high-strength concretes (HSCs), which generally possess a higher overall density and lower permeability [7, 27]. The increase in the popularity and use of these high-strength concretes in building construction heightens the urgency of developing better methods for predicting and mitigating the effects of this spalling behaviour [1].

Fibre-reinforced concretes have been used in building construction since at least the 1970s [30]. Fibres have generally been used in these concrete mixes to improve the ductility and durability of the material [13, 33]. Polymer fibres have also been shown, however, to significantly contribute to spall resistance within concrete components during fire [8, 9].

The exact mechanisms causing this increased spalling resistance are not yet fully understood, but it is thought that the fibres facilitate moisture transport during fire exposure, which in turn contributes to relief of internal stresses in the spalling-susceptible region [9, 25]. The formation of a fibre-induced micro-cracking network has been identified as one important aspect of this process [25]. At the present time, there is also no comprehensive understanding of how fibre properties, such as polymer type, diameter, length, shape, and density, affect overall spalling performance of concrete members during fire. Optimization of these parameters will need to be completed before polymer-fibre reinforced concretes are adopted into widespread use for fire protection.

1.2 X-Ray Computed Tomography

X-ray computed tomography (CT) has been used in non-destructive concrete research applications for more than 30 years [15, 19]. In this scanning method, a sample is placed on a rotating table between an X-ray source and an X-ray detector [4]. This causes an X-ray attenuation image of the sample to be projected upon the detector. By recording these projected images during the 360° rotation of the sample, mathematically-based reconstruction algorithms can be used to produce a three dimensional representation of X-ray attenuation within the sample [3]. The X-ray

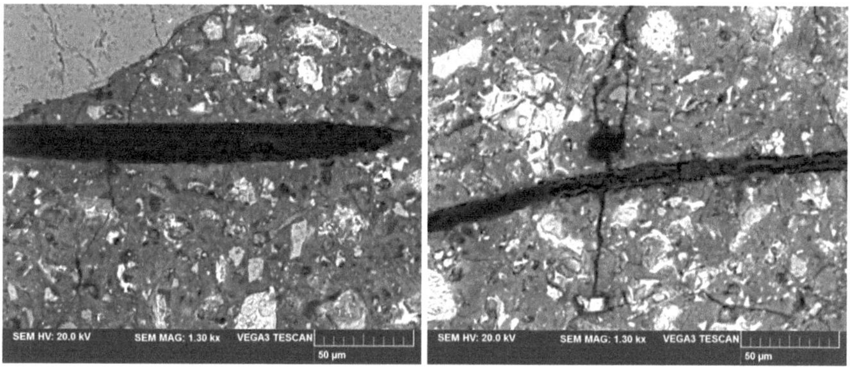

Fig. 1 Scanning Electron Microscope (SEM) images of polymer fibres before (left) and after (right) heating to 300 °C

attenuation, which is roughly correlated with density, of individual component materials within the sample can then be identified and objects made from these component materials can be individually separated and analysed.

1.3 Objectives and Significance

Previous research has shown that small cracks develop in the mortar that surrounds the fibres during curing (Fig. 1). One goal of this research was to investigate to what extent these curing cracks contribute to overall material permeability prior to and during heating.

Another goal of this research was to identify the presence of correlations between fibre and cracking orientation characteristics and to quantitatively define the strength of those correlations. Previous research has demonstrated that the fibre fields in FRCs tend to be highly anisotropic because of flow during the casting process [5, 6, 17, 18, 20–22, 26]. It is thought that this anisotropy could have a major effect on fire resistance and, if properly controlled, may serve as a means of significantly improving spalling resistance.

The third goal of this research was to quantitatively measure the crack growth during incremental heating through the use of X-ray computed tomography (CT). This CT-based data would be particularly well suited for the calibration and validation of computational models of the spalling phenomenon.

2 Materials and Sample Preparation

The investigated specimens were made of HSC and were reinforced with different amounts and types of polypropylene (PP) fibres (Table 1). The mixture HSC / PPa2 contained 2 kg/m^3 conventional PP-fibres. In contrast, for HSC / PPb1 only 1 kg/m^3 pre-treated PP-fibres was added. The PPb-fibres were pre-treated during the manufacturing process using electron irradiation. This leads to a decreased fibre-melt viscosity [24] and improves the fire performance of the concrete, despite the reduced amount of added PP-fibres.

Initially, cubes (100 × 100 × 100 mm^3) were cast for both mixtures. After demoulding on the next day, these cubes were stored under water for six days and subsequently in a climate chamber at 20 °C and 65 % relative humidity for a minimum of 83 days. Subsequently, cylindrical specimens with diameters of 12 mm and lengths of 100 mm were extracted from the cubes. A single cylindrical sample from each of the two cubes was selected for heating and CT analysis. These two samples will henceforward be referred to as sample PPb1 and sample PPa2, in reference to their composition. Before the first CT-measurement was carried out, these cylinders were fixed firmly within a customized set of mobile clamps, which ensured consistent mounting positions during repetitive CT measurements.

Table 1 HSC mixture

<table>
<tr><th colspan="3" rowspan="2">Component</th><th colspan="2">Content [kg/m³] in concrete type</th></tr>
<tr><th>HSC / PPa-2</th><th>HSC / PPb-1</th></tr>
<tr><td colspan="3">Cement CEM 42,5 R</td><td colspan="2">580</td></tr>
<tr><td colspan="3">Water (w/c = 0,33)</td><td colspan="2">173</td></tr>
<tr><td rowspan="3">Aggregate (quarzitic)</td><td>0/2</td><td rowspan="3">[mm]</td><td colspan="2">744</td></tr>
<tr><td>2/4</td><td colspan="2">223</td></tr>
<tr><td>4/8</td><td colspan="2">521</td></tr>
<tr><td colspan="3">Silica fume</td><td colspan="2">63,8</td></tr>
<tr><td colspan="3">Superplasticizer</td><td colspan="2">27,8</td></tr>
<tr><td colspan="2" rowspan="2">PP-fibre type (∅ 15,4 µm, L = 6 mm)</td><td>a</td><td>2</td><td>-</td></tr>
<tr><td>b</td><td>-</td><td>1</td></tr>
</table>

Where:
PPa = conventional polypropylene fibres
PPb = pre-treated polypropylene fibres

3 Test Methods

After preparing and mounting the concrete cylinders, a CT-measurement was completed. Thereby, the initial structure of the concrete was analysed. Afterwards, the clamped concrete cylinders were heated using a special heating regime to reach specific maximum temperatures (see Thermal Loading). Subsequent to each heating/cooling phase, a further CT-measurement was carried out to investigate the crack formation in the fibre-reinforced HSC as a result of the thermal exposure.

3.1 Thermal Loading

During the test series, the specimens were heated to certain target temperatures (150, 160, 170, 180, 190, 200, 250, 300 °C) using an electrical furnace (Fig. 2). During thermal testing, the heat was increased at a rate of 1 K/min until the respective target temperature was reached. The target temperature was then sustained for one hour, followed by cooling at a maximum rate of −0,5 K/min to room temperature prior to CT-scanning.

The used electrical furnace is controlled by a temperature controller. The oven temperature is determined on the basis of the temperature measurement of a thermocouple, which is fixed in the rear wall of the furnace. Figure 3 shows an example temperature curve measured with an additional installed thermocouple near the

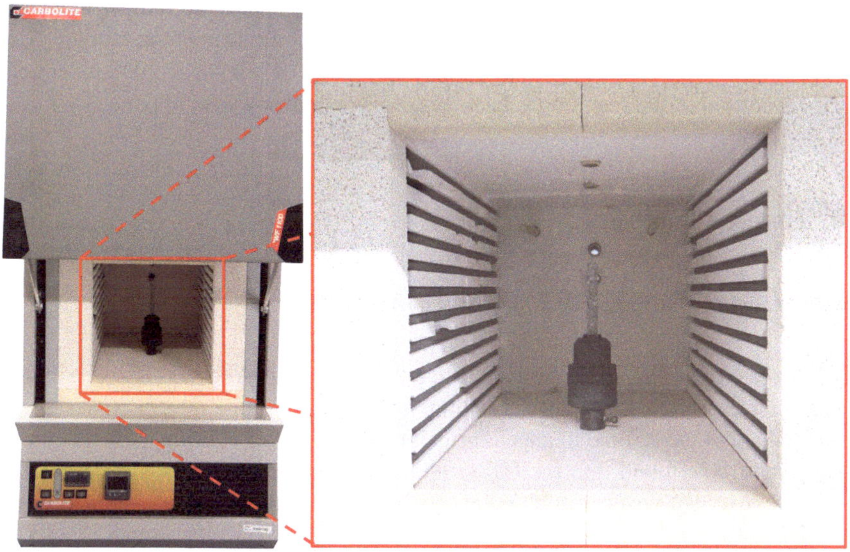

Fig. 2 Electrical furnace containing a typical clamped specimen

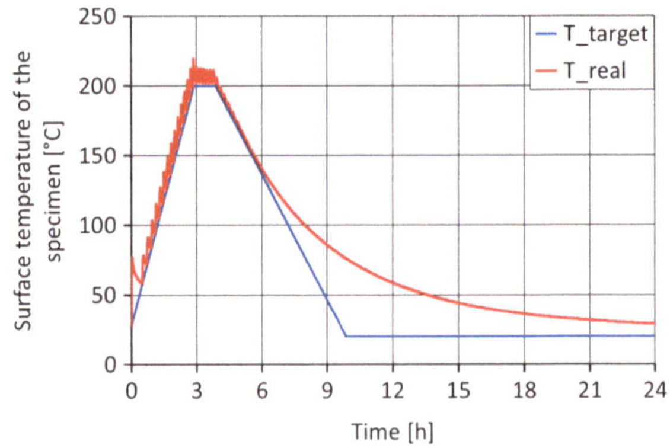

Fig. 3 Example temperature curve for heating test at 200 °C

specimen in comparison with the target temperature curve. Thereby it can be seen that the temperature in the furnace is controlled very well during heating phase. During the cooling phase the actual cooling rate is slower than the prescribed one because of the nonlinearity of the natural cooling process.

3.2 CT Scanning

During this research program, an acceleration voltage of 60 kV and current of 130 μA were used for the X-ray source. The X-ray beam was also filtered using a 1 mm thick Aluminium plate immediately upon leaving the source in order to remove unwanted bandwidths from the X-ray beam and, in turn, make the resulting images more clear. The flat panel detector used for this scanning contained a 2048 × 2048 pixel field. The resolution of the resulting CT images of the samples was 6.18 μm.

4 Image Processing

4.1 Initial Processing Procedures

All image processing was completed using the program MATLAB [16]. The images were first corrected for beam hardening, which is a CT phenomenon causing the outer edges of the sample to appear brighter than its centre. During subsequent data analysis, it was found that the full sized sample image, which was 2048 × 2048 × 2048 voxels (a voxel is a 3D pixel), was much too large for the available image processing algorithms and computer system. Thus, a cubic sub-volume of 1200 × 1200 × 1200 voxels was digitally extracted from the centre of the original image and used for all subsequent analysis. Through the use of this sub-volume, all resulting computational demands and run times were reduced by a factor of approximately five times.

To enable the density and orientation analysis of the fibres and cracks, these materials first needed to be identified and separated within the images. Although individual component materials can generally be separated within CT images using grayscale segmentation methods that was not sufficient for these sample images. The reason for this complication was that the X-ray attenuation levels of the air pores, the cracks, and the polymer fibres were found to all be very similar and partially overlapping. Thus, more complex methods of fibre and crack detection needed to be developed.

4.2 *Fibre Identification*

Initially, template matching methods were used in an attempt to identify the polymer fibres, but with only limited success. One reason for the failure of this method may have been that the fibres had such a small size within the images (only approximately 2.5 voxels in diameter) that their shape was not sufficiently well-defined to consistently match the template characteristics. Another problem with this method was that the polymer fibres often exhibited significant bending within the material, which made it even more difficult to define and match a consistent fibre shape.

To overcome these limitations, a customized multi-step approach was developed for fibre identification that exhibited considerable versatility and accuracy. The individual steps of this analysis procedure will be outlined in the description below:

Step 1. The triangle segmentation method [34, 35] was used to identify a boundary between low attenuation elements within the sample, including air pores, cracks, and polymer fibres, and high attenuation elements, including aggregate and mortar, on the voxel intensity histogram (Fig. 4). Using this attenuation threshold, a sample image containing only air pores, cracks, and fibres could be created and used for further analysis (Fig. 5).

Step 2. All objects within the low-attenuation image were individually analysed and those with a volume less than 16% of the standard individual fibre volume were removed from the image (Fig. 5). This was done in order to eliminate both noise and micro-pores from the image. The reason that some objects with less than 100% of the fibre volume were retained within the image was that, since the analysis was conducted upon a cubic sub-volume, many partial fibres existed along the image edge. It was desirable to retain these partial fibre segments for subsequent analysis steps as they were expected to contribute significantly to overall sample performance. The volume limitation of 16% was found through trial and error to produce a good balance between elimination of unwanted objects and retention of partial fibre segments.

Step 3. The objects within the image were contracted and then subsequently dilated by an amount equivalent to the fibre radius (rounded upward). This resulted in an image containing only objects larger in diameter than fibres. These macro-pores

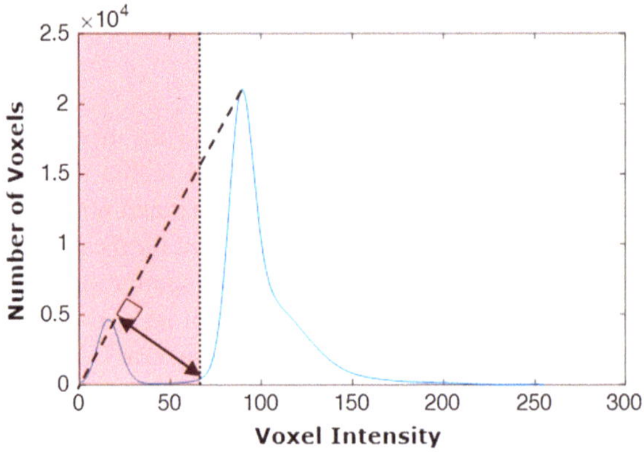

Fig. 4 Definition of low-high attenuation threshold on the voxel intensity histogram

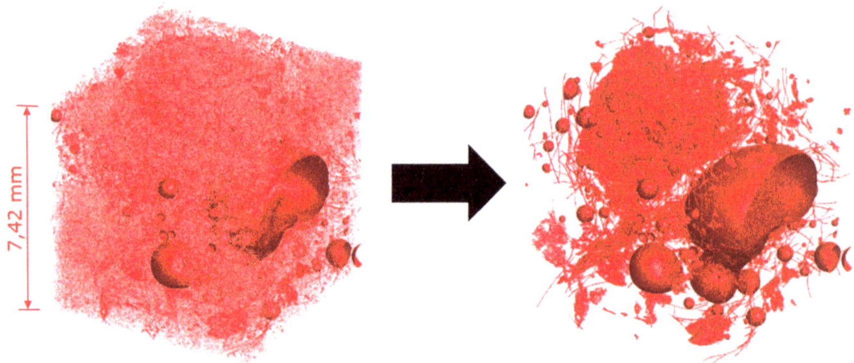

Fig. 5 Image of low-attenuation materials in sample PPb1 after step 1 (left) and step 2 (right)

were subsequently removed from the image produced by step 2 in order to reduce the number of non-fibre objects (Fig. 6).

Through the removal of all objects with diameter larger than and volume smaller than that of the fibres in steps 2 and 3, it was originally thought that a clear, fibre-only image would result. It was soon found, however, that the samples contained may small micro-cracks of width similar to the fibre diameter even prior to the application of heat. Closer inspection revealed that these micro-cracks were almost exclusively present within the aggregate, rather than the mortar.

Step 4. Since the fibres, in contrast to the cracks, are only present within the mortar, and never within aggregates, it was possible to develop an algorithm for separating the fibres from the remaining cracks. In this algorithm, each object was dilated by an amount equivalent to the fibre radius (rounded upward) and the attenuation of this dilated region was analysed. If the attenuation of the dilated region around an individual object corresponded to that representative of mortar (Figs. 7 and 8), the

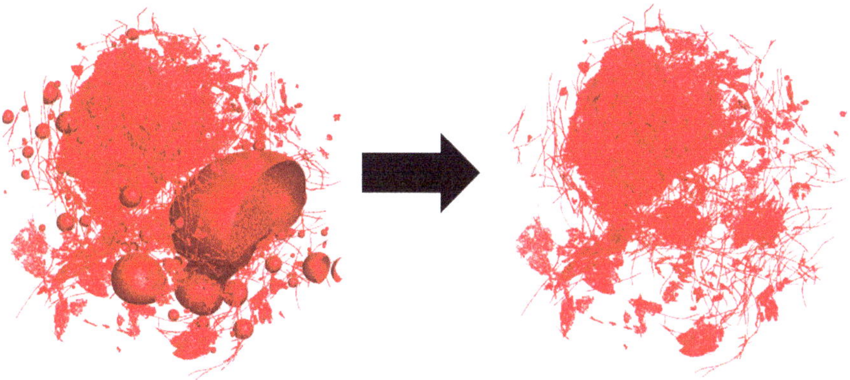

Fig. 6 Image of low-attenuation materials in sample PPb1 after step 2 (left) and after step 3 (right)

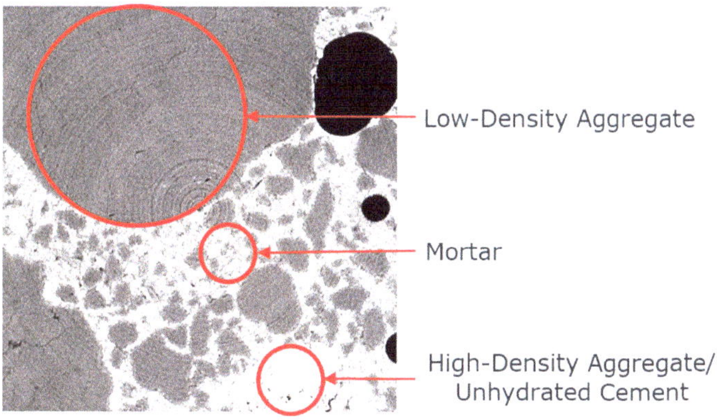

Fig. 7 Attenuation image of the fibre-reinforced concrete material visually demonstrating the attenuation differences between aggregate and mortar

object was considered a fibre and retained. All other objects were considered as micro-cracks within the aggregate and eliminated (Fig. 9).

The resulting fibre images for samples PPb1 and PPa2 can be compared in Fig. 10. Although these images still contain some non-fibre objects, the overall contribution of these objects to the measured fibre characteristics is assumed to be small. These non-fibre objects are thought to primarily consist of cracks or small voids that intersect with fibres as well as ring artefacts.

As previously described, the sample PPa2 was fabricated to contain a fibre volume twice that of sample PPb1. The differences in fibre volume between the analysed sub-volumes of the two samples shown in Fig. 10, however, exhibit a much larger volume difference. The fibre volume percentage within each analysed sub-volume could be calculated by summing the number of voxels of fibre material and dividing that number by the total number of voxels in the image. Using this method a fibre-

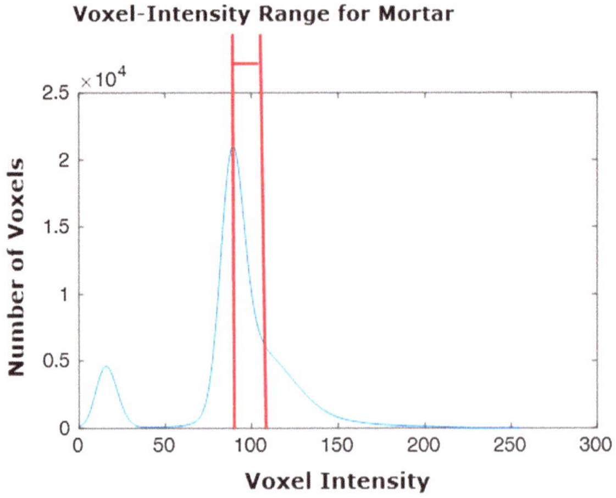

Fig. 8 Definition of attenuation threshold boundaries for mortar material on the voxel intensity histogram

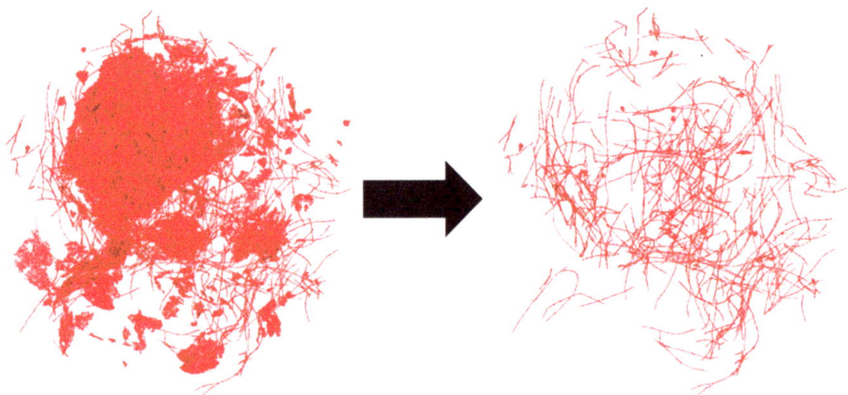

Fig. 9 Image of low-attenuation materials in sample PPb1 after step 3 (left) and after step 4 (right)

volume percentage of 0.202% was measured for the PPa2 sub-volume, which was very close to the 0.22% fibre-volume percentage used in the mix design.

For the PPb1 sub-volume, however, a fibre-volume percentage of only 0.044% was measured as opposed to the 0.11% fibre-volume percentage used in the mix design. There are many possible reasons for such a discrepancy. The most likely causes are inhomogeneity of fibre distribution within the sample and the presence of large aggregates or air-voids within the sub-volume selected for analysis. Both of these sources of error would be significantly reduced through the selection of larger sub-volumes for analysis. There is, thus, an impetus for further development of these image processing algorithms in order to reduce their computational requirements and enable the analysis of larger sub-volumes in future scanning and analysis efforts.

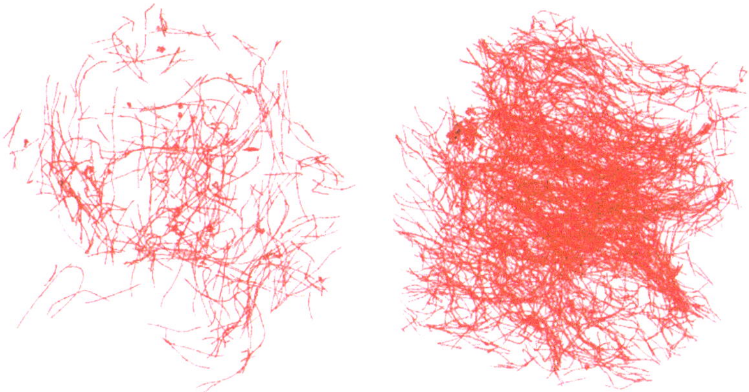

Fig. 10 Fibre images for samples PPb1 (left) and PPa2 (right)

4.3 Fibre Analysis

Once the fibres had been identified and isolated within the images, they were then analysed for both density and orientation characteristics. The orientation analysis was completed using the Hessian-based method [6, 14]. In this method, the grayscale images resulting from CT are considered to be three-dimensional functions that are twice differentiable in all directions [10]. By calculating the Hessian matrix at a given voxel within a fibre, partial second derivatives can be computed:

$$H = \begin{bmatrix} \frac{\partial^2 I}{\partial x^2} & \frac{\partial^2 I}{\partial x \partial y} & \frac{\partial^2 I}{\partial x \partial z} \\ \frac{\partial^2 I}{\partial y \partial x} & \frac{\partial^2 I}{\partial y^2} & \frac{\partial^2 I}{\partial y \partial z} \\ \frac{\partial^2 I}{\partial z \partial x} & \frac{\partial^2 I}{\partial z \partial y} & \frac{\partial^2 I}{\partial z^2} \end{bmatrix} \quad (1)$$

with H = Hessian Matrix and I = Grayscale Sample Image Matrix.

At this point, the second derivative in the direction of the longitudinal axis of the fibre will be much less than those in the transverse directions. The orientation of fibres can, thus, be assessed by computing the eigenvalues and eigenvectors of the Hessian matrix at each voxel within a fibre (Fig. 11). The primary fibre orientation recorded at a single fibre voxel, therefore, is the eigenvector corresponding to the smallest eigenvalue [29].

For the fibre density analysis method, a small cell size was selected, comprising a cube with 120 voxel (742 μm) long sides. The fibre image was broken up into an array of these cells, and the number of white (i.e., fibre) voxels in each was counted. By dividing the number of fibre voxels in each cell by the total number of cell voxels, estimates of local fibre density within the sample could be obtained.

The cell size used for the density analysis was selected such that the length of each cube side represented about a 10th of the overall length of a typical image array side.

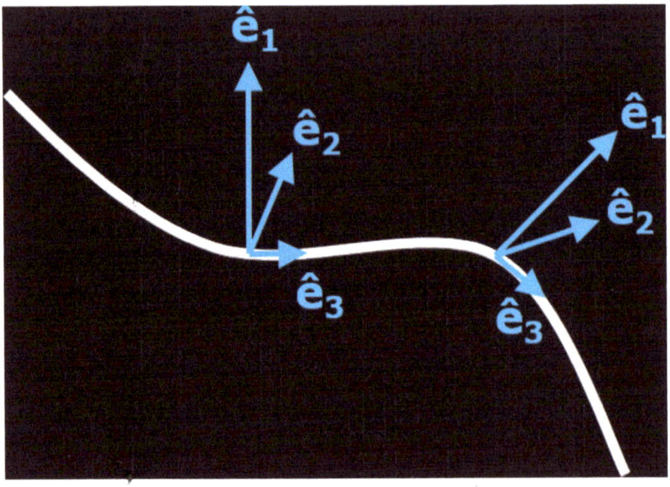

Fig. 11 Eigenvectors of the Hessian matrix at two points within a fibre

This meant that a sufficient number of cells (1000) would be available for analysis to enable meaningful statistical evaluation. This is because for a pseudo-random phenomenon, such as fibre density variation, a large number of samples need to be collected to observe meaningful trends within the statistical data. At the same time, however, there was a desire to avoid cell sizes that were too small because they might be too highly influenced by the presence of single fibres.

4.4 *Cracking Analysis*

The lack of a consistent size or shape for the crack geometry also made template matching techniques of limited use for crack identification. These characteristics made the use of a customized isolation approach similar to that used for the fibres unsuitable as well.

Past research has demonstrated, however, that many cracking characteristics can be measured through the observation of changes in void properties in images of progressively damaged samples [21, 23]. In this approach, all void characteristics, including volume, surface area, and orientation, within the initial, undamaged scan of a sample are assumed to be related to entrained and entrapped air. Thus, any change in these void characteristics, such as volume and surface area growth, seen in later scans of damaged samples can be assumed to be due to cracking and, thus, represented as crack characteristics.

It is important to note, however, that when analysing the orientation of cracks, a modified version of the approach used for fibre orientation analysis must be applied. Since cracks are planar objects, they cannot be characterized by a single, parallel vector in the way that fibres can. Rather, the cracks must be characterised by a vector which is perpendicular to the plane of the crack. Thus, in the Hessian characterization of crack orientation, the orientation angle represented by the eigenvector corresponding to the largest eigenvalue of the Hessian matrix must be used.

5 Results and Discussion

5.1 Results of Fibre Analysis

Various methods exist to depict and analyse orientation data in three dimensions. For the present analysis, coordinates have been converted from a Cartesian to a spherical system [32]:

$$r = \sqrt{x^2 + y^2 + z^2}\,, \tag{2}$$

$$\theta = \tan^{-1}\left(\frac{y}{x}\right)\,, \tag{3}$$

$$\phi = \cos^{-1}\left(\frac{z}{r}\right)\,. \tag{4}$$

In this spherical coordinate system, orientations are characterized by angles θ and ϕ (Fig. 12). The angle θ represents the azimuthal angle in the x-y plane from the x-axis (in this context the cylindrical axis of the sample is denoted as the z-axis), with $0 < \theta < 360°$ (Eq. 3). The angle ϕ represents the polar angle from the positive z-axis. Since the fibres are fully symmetric, a symmetry condition is likewise imposed on ϕ, with $0 < \phi < 90°$ (Eq. 4).

Fig. 12 Spherical coordinate system

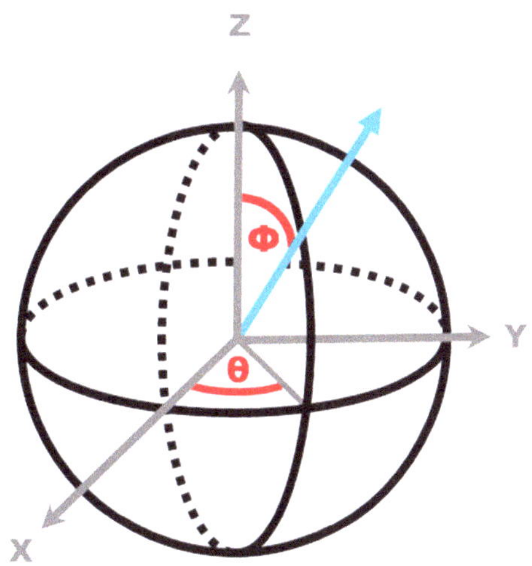

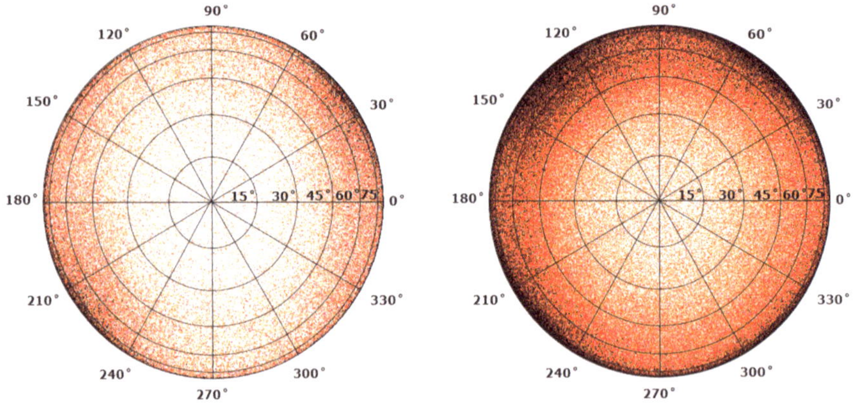

Fig. 13 Orthogonal projections of fibre orientation for samples PPb1 (left) and PPa2 (right). Projection radius: $0 < \phi < 90°$; projection circumference: $0 < \theta < 360°$

Figure 13 provides orthogonal projections of the fibre orientation data for samples PPb1 and PPa2. The depiction of three-dimensional data on a two-dimensional plane makes these projections unsuitable for evaluating isotropy along the ϕ-axis but the projections can be used very effectively for evaluating isotropy along the θ-axis. For sample PPb1, there appeared to be an anisotropic orientation of fibres along roughly the $\theta = 30°/210°$ axis. A similar, although less clear anisotropy also appeared in the projection for sample PPa2. The darker colours of the PPa2 projection were caused by the higher fibre-volume content already discussed.

Fibre density results can be displayed in the form of histograms (Figs. 14 and 15). For clarity, the y-axis in these histograms has been depicted in logarithmic scale. This is necessary since the vast majority of cubic sub-volumes contain no fibres or only a few fibre voxels. These histograms confirm the higher fibre-volume content of sample PPa2.

5.2 *Results of Cracking Analysis*

The void data were also analysed using spherical coordinates and changes in void orientation characteristics, which correspond to crack orientation characteristics, can be depicted using orthogonal projections (Fig. 16).

Recalling that the crack orientations were measured in terms of orientation normal to the cracking plane, it can be observed that for both PPb1 and PPa2, there appears to be an anisotropy along roughly the $\theta = 130°/310°$ plane. This indicates that the primary orientation of the cracks is parallel to the primary fibre orientation. This is because the primary normal orientation of the cracks (depicted in Fig. 16) is roughly orthogonal to the primary fibre orientation (depicted in Fig. 13), which is typical for a plane running parallel to a line.

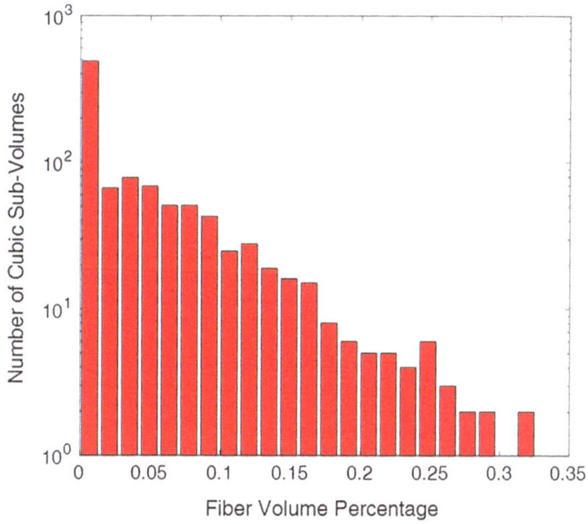

Fig. 14 Histogram of fibre density for sample PPb1

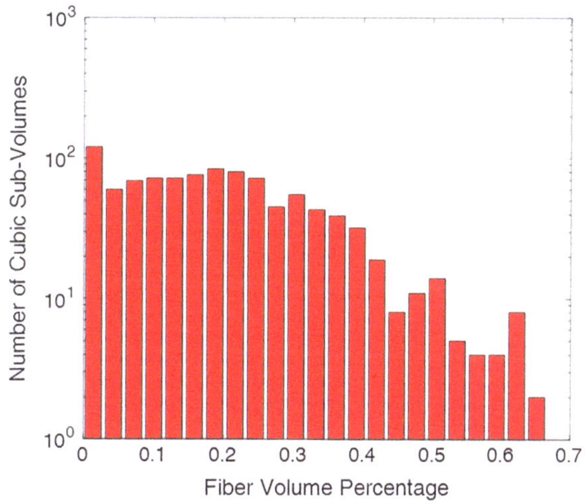

Fig. 15 Histogram of fibre density for sample PPa2

5.3 Comparison of Fibre and Cracking Characteristics

Comparison of fibre and cracking characteristics based purely on Figs. 13 and 16 is rather qualitative and difficult, however. In order to provide a more amenable means for direct comparison, histograms of fibre and crack-normal orientation were created and overlaid (Figs. 17 and 18).

The relationship between the θ-orientations of the fibres and the normal vectors to the cracking plane for PPb1 exhibit behaviour typical of cracks running along the primary fibre orientation direction, with very few fibres oriented along the directions normal to the cracking plane. This correlation is less clear for the PPa2 data.

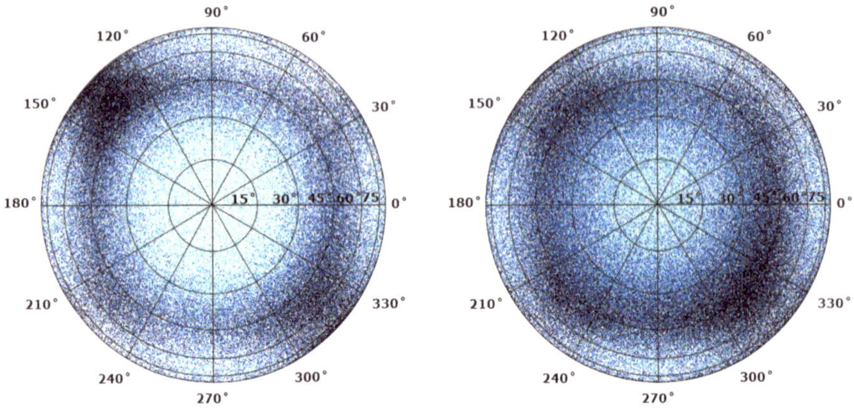

Fig. 16 Orthogonal projections of the normal orientation vectors measured for the cracks in samples PPb1 (left) and PPa2 (right) after 300 °C heating. Projection radius: $0 < \phi < 90°$; projection circumference: $0 < \theta < 360°$

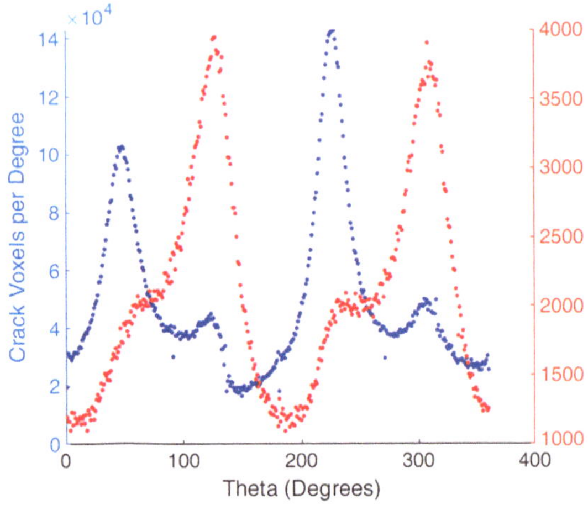

Fig. 17 Relationship between fibre (red) and crack-normal (blue) θ-orientation for sample PPb1

The reasons for this are unclear, but may be related to the higher volume of fibres contained within this sample. This increase in fibre volume may lead to more concerted behaviour of fibres through their combination in bundles as opposed to the less dense fibres in PPb1, which are more likely to interact with the concrete material individually. The presence of crack or ring artefacts within the fibre images could also contribute to this inconsistency.

Fibre-cracking histograms have also been created relative to the ϕ-orientation (Figs. 19 and 20). Direct evaluation of these histograms is complicated by the fact that fibre density is not uniform along the ϕ-axis due to exponential growth of volume in the unit sphere corresponding with linear increases in ϕ-angle. Very similar trends in ϕ-orientation are observable for the fibre and crack-normal data of both samples

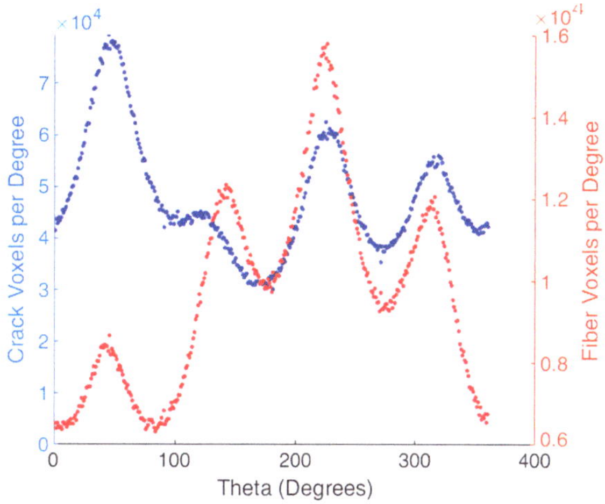

Fig. 18 Relationship between fibre (red) and crack-normal (blue) θ-orientation for sample PPa2

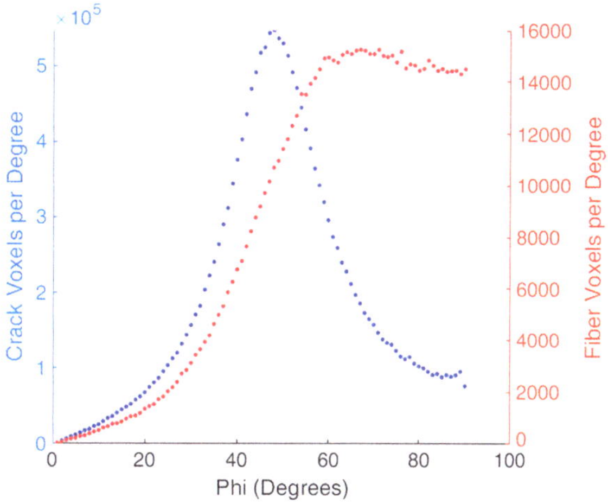

Fig. 19 Relationship between fibre (red) and crack-normal (blue) ϕ-orientation for sample PPb1

PPb1 and PPa2. These figures provide confirmation that an offset between the fibre and crack-normal orientation exists. Although not clearly orthogonal to one another, as would be expected for perfectly parallel cracks and fibres, this orientation offset does indicate that the cracks tend to orient themselves more strongly parallel to fibres than perpendicular to them.

A detailed analysis of fiber-crack interactions within localized regions of these samples can also be found in [31].

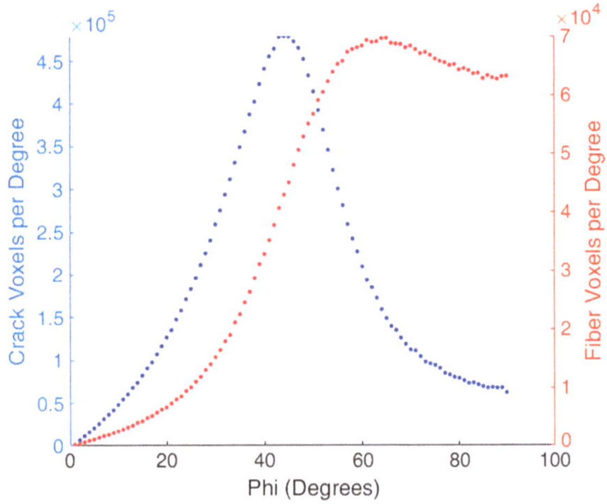

Fig. 20 Relationship between fibre (red) and crack-normal (blue) ϕ-orientation for sample PPa2

5.4 *Correlation between Heating and Cracking Characteristics*

Much clearer trends can be observed in the growth of cracking relative to temperature increase. Figures 21, 22, 23 and 24 depict the growth of cracking relative to applied heating for both samples PPb1 and PPa2. In these figures, two different measurements are used to evaluate cracking growth. Figures 21 and 22 are plotted relative to crack volume while Figs. 23 and 24 are plotted relative to crack surface area.

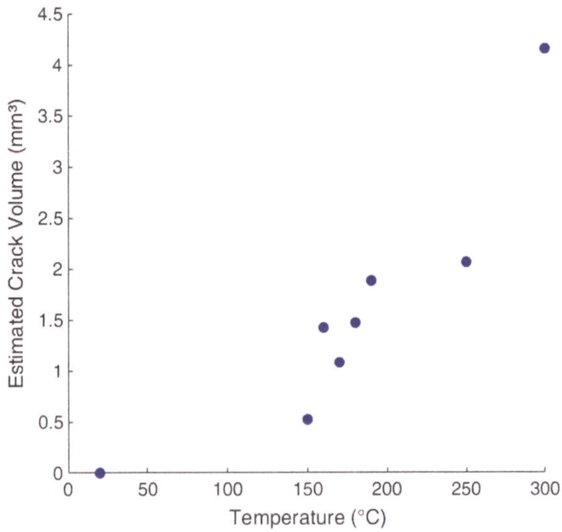

Fig. 21 Relationship between temperature and crack volume for sample PPb1

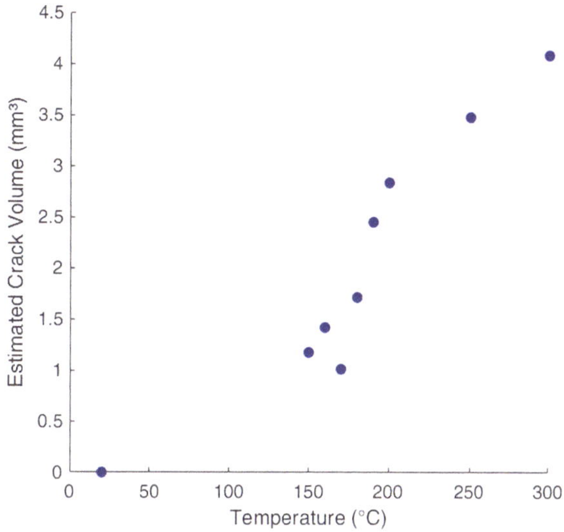

Fig. 22 Relationship between temperature and crack volume for sample PPa2

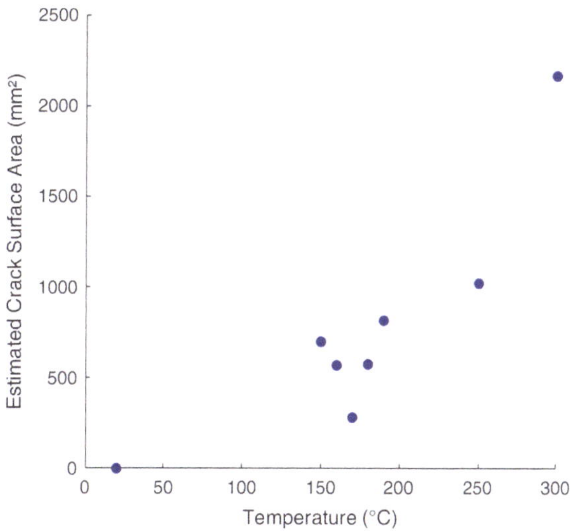

Fig. 23 Relationship between temperature and crack surface area for sample PPb1

In each of these figures, clear trends are visible between cracking and heating characteristics. These quantitative trends offer great promise for the calibration and validation of fire-damage models within finite element analysis codes for concrete. Basic fracture mechanics uses crack surface area as one parameter for calculating fracture energy [2]. This may make crack surface area the optimal damage parameter for use in numerical modelling.

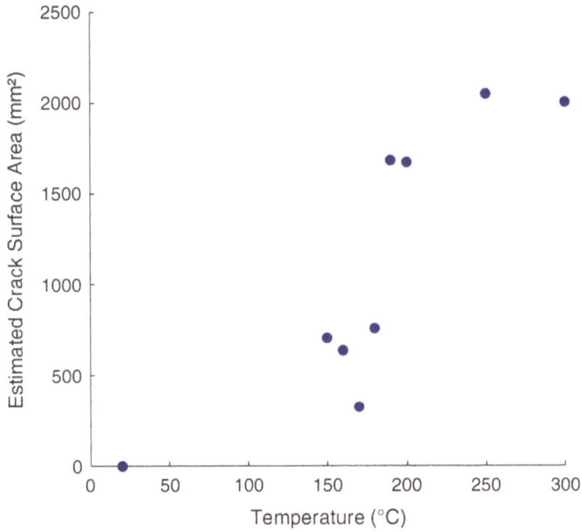

Fig. 24 Relationship between temperature and crack surface area for sample PPa2

Crack volumes have, however, also previously been shown to follow clear trends relative to work-of-load and stiffness reduction [11, 12]. The optimal numerical approach might be to develop a calibration and validation approach that combines the measurements of both of these cracking characteristics.

6 Conclusions

The results of this research effort have demonstrated that the orientation of both fibres and cracks in polymer-fibre reinforced concrete tend to be anisotropic. It is thought that this anisotropy is predominantly influenced by casting method, but can also be affected by the presence of large aggregates or voids within the sample. The results of the orientation analysis also indicated that crack geometry characteristics may be correlated with fibre orientation, with cracks tending to run parallel to fibre beds. This could have major implications for structural level performance since it would indicate that fire resistance may be related to casting method.

Clear quantitative relationships were also observed between heating and increasing cracking levels, expressed in terms of both crack surface area and crack volume. These relationships can serve as the basis for calibration and validation of finite element models used for simulating heat-related spalling behaviour.

7 Future Work

Future research is needed both to improve the accuracy and reliability of the analysis procedures used in this research and to develop new analysis procedures for evaluating phenomena of relevance to the spalling behaviour of concrete. One reason that only aggregated cracking and fibre characteristics could be measured in this research was the lack of accurate digital volume correlation (DVC) tools. The development and application of such tools would enable direct comparison, including subtraction, of individual image features among multiple images in a heating or loading series. This would not only make it easier to distinguish crack growth and fibre failure, but it would also enable more accurate measurements of crack-fibre property correlation. Up to the present time it has been very difficult to develop accurate DVC tools for concrete analysis because of the cracking discontinuities typical of concrete failure, which are generally more difficult for DVC methods to accommodate than the simple strains typical of plastically deforming materials.

New methods also need to be developed for assessing the cross-linking of fibre beds during heat-related cracking. Although this phenomenon has been observed using SEM, it has been difficult to develop a method that accurately quantifies the material behaviour. This is further complicated by the fact that many of the connecting cracks observed during SEM scanning appear to be below the resolution of most laboratory CT systems. It is believed that significant progress could be made in measuring this phenomenon through the skilled application of synchrotron-CT scanning in combination with DVC image processing tools.

Finally, further development of the image processing algorithms detailed in this paper is needed. These algorithms should be streamlined to enable orientation analysis of larger regions of interest within samples, which would minimize the magnitude of error introduced by individual material features, such as single stones or voids. Further research is also needed to improve the precision of the fibre-cracking orientation comparison and to statistically quantify the correlation level. Only through the accurate measurement of statistical correlation between fibre and cracking orientation can firm conclusions be drawn about the optimal casting procedures for the construction of fire-resistant building components.

References

1. Aitcin, P.C.: High Performance Concrete. CRC Press (1998). https://doi.org/10.1201/9781420022636
2. Bazant, Z.P., Planas, J.: Fracture and Size Effect in Concrete and Other Quasibrittle Materials, vol. 16. CRC Press (1997)
3. Feldkamp, L.A., Davis, L.C., Kress, J.W.: Practical cone-beam algorithm. J. Opt. Soc. Am. A **1**(6), 612–619 (1984). https://doi.org/10.1364/JOSAA.1.000612
4. Flannery, B.P., Deckman, H.W., Roberge, W.G., D'Amico, K.L.: Three-dimensional x-ray microtomography. Science **237**(4821), 1439–1444 (1987). https://doi.org/10.1126/science.237.4821.1439

5. Herrmann, H., Lees, A.: On the influence of the rheological boundary conditions on the fibre orientations in the production of steel fibre reinforced concrete elements. Proc. Est. Acad. Sci. **65**(4), 408–413 (2016). https://doi.org/10.3176/proc.2016.4.08.Open-AccessCC-BY-NC4.0
6. Herrmann, H., Pastorelli, E., Kallonen, A., Suuronen, J.P.: Methods for fibre orientation analysis of x-ray tomography images of steel fibre reinforced concrete (SFRC). J. Mater. Sci. **51**(8), 3772–3783 (2016). https://doi.org/10.1007/s10853-015-9695-4
7. Hertz, K.: Explosion of silica-fume concrete. Fire Saf. J. **8**(1), 77 (1984). https://doi.org/10.1016/0379-7112(84)90057-2
8. Jansson, R.: Material Properties Related to Fire Spalling of Concrete. Division of Building Materials. Lund Institute of Technology, Lund University (2008)
9. Jansson, R.: Fire spalling of concrete: theoretical and experimental studies. Ph.D. thesis, KTH Royal Institute of Technology (2013)
10. Krause, M., Hausherr, J.M., Burgeth, B., Herrmann, C., Krenkel, W.: Determination of the fibre orientation in composites using the structure tensor and local x-ray transform. J. Mater. Sci. **45**(4), 888 (2010). https://doi.org/10.1007/s10853-009-4016-4
11. Landis, E.N.: Toward a physical damage variable for concrete. J. Eng. Mech. **132**(7), 771–774 (2006). https://doi.org/10.1061/(ASCE)0733-9399(2006)132:7(771)
12. Landis, E.N., Zhang, T., Nagy, E.N., Nagy, G., Franklin, W.R.: Cracking, damage and fracture in four dimensions. Mater. Struct. **40**(4), 357–364 (2007). https://doi.org/10.1617/s11527-006-9145-5
13. Li, V.C., Wang, S.: Microstructure variability and macroscopic composite properties of high performance fiber reinforced cementitious composites. Probab. Eng. Mech. **21**(3), 201–206 (2006). https://doi.org/10.1016/j.probengmech.2005.10.008 (Probability and Materials: from Nano- to Macro-Scale)
14. Lorenz, C., Carlsen, I.C., Buzug, T.M., Fassnacht, C., Weese, J.: Multi-scale line segmentation with automatic estimation of width, contrast and tangential direction in 2D and 3D medical images. In: CVRMed-MRCAS'97, pp. 233–242. Springer (1997)
15. Martz, H.E., Scheberk, D.J., Roberson, G.P., Monteiro, P.J.: Computerized tomography analysis of reinforced concrete. Mater. J. **90**(3), 259–264 (1993)
16. Mathworks T: Matlab. r2014a. Natick. MA, USA (2016)
17. Mishurova, T., Léonard, F., Oesch, T., Meinel, D., Bruno, G., Rachmatulin, N., Fontana, P., Sevostianov, I.: Evaluation of fiber orientation in a composite and its effect on material behavior. In: Proceedings of the 7th Conference on Industrial Computed Tomography (ICT) held February 7–9, 2017, Leuven, Belgium, vol. 22(03). NDT.net (2017). http://www.ndt.net/?id=20818
18. Mishurova, T., Rachmatulin, N., Fontana, P., Oesch, T., Bruno, G., Radi, E., Sevostianov, I.: Evaluation of the probability density of inhomogeneous fiber orientations by computed tomography and its application to the calculation of the effective properties of a fiber-reinforced composite. Int. J. Eng. Sci. **122**, 14–29 (2018). https://doi.org/10.1016/j.ijengsci.2017.10.002
19. Morgan, I., Ellinger, H., Klinksiek, R., Thompson, J.N.: Examination of concrete by computerized tomography. J. Proc. **77**(1), 23–27 (1980)
20. Oesch, T., Landis, E., Kuchma, D.: A methodology for quantifying the impact of casting procedure on anisotropy in fiber-reinforced concrete using x-ray ct. Mater. Struct. **51**(3), Article 73, 1–13 (2018). https://doi.org/10.1617/s11527-018-1198-8
21. Oesch, T.S.: Investigation of fiber and cracking behavior for conventional and ultra-high performance concretes using x-ray computed tomography. University of Illinois at Urbana-Champaign (2015)
22. Oesch, T.S.: In-situ ct investigation of pull-out failure for reinforcing bars embedded in conventional and high-performance concretes. In: Proceedings of 6th Conference on Industrial Computed Tomography (ICT), vol. 21 (2016)
23. Oesch, T.S., Landis, E.N., Kuchma, D.A.: Conventional concrete and UHPC performance–damage relationships identified using computed tomography. J. Eng. Mech. **142**(12), 04016101 (2016)
24. Pistol, K.: Wirkungsweise von polypropylen-fasern in brandbeanspruchtem hochleistungsbeton. doctoralthesis, Bundesanstalt für Materialforschung und -prüfung (BAM) (2016)

25. Pistol, K., Weise, F., Meng, B., Schneider, U.: The mode of action of polypropylene fibres in high performance concrete at high temperatures. In: 2nd International RILEM Workshop on Concrete Spalling due to Fire Exposure, pp. 289–296. RILEM Publications SARL (2011)
26. Pujadas, P., Blanco, A., Cavalaro, S., de la Fuente, A., Aguado, A.: Fibre distribution in macro-plastic fibre reinforced concrete slab-panels. Constr. Build. Mater. **64**, 496–503 (2014)
27. Sanjayan, G., Stocks, L.: Spalling of high-strength silica fume concrete in fire. Mater. J. **90**(2), 170–173 (1993)
28. Stelzner, L., Powierza, B., Weise, F., Oesch, T.S., Dlugosch, R., Meng, B.: Analysis of moisture transport in unilateral-heated dense high-strength concrete. In: Proceedings from the 5th International Workshop on Concrete Spalling, pp. 227–239 (2017)
29. Trainor, K.: 3-D analysis of energy dissipation mechanisms in steel fiber reinforced reactive powder concrete. Master's thesis, The University of Main (2011)
30. Urbana-Champaign UoIa: The history of concrete: a timeline. Department of Materials Science and Engineering. http://matse1.matse.illinois.edu/concrete/hist.html (2015)
31. Weise, F., Stelzner, L., Weinberger, J., Oesch, T.S.: Influence of the pre-treatment of pp-fibres by means of electron irradiation on the spalling behaviour of high strength concrete. In: Proceedings from the 5th International Workshop on Concrete Spalling, pp. 345–358 (2017)
32. Weisstein, E.W.: Spherical Coordinates. From MathWorld–A Wolfram Web Resource (2017). http://mathworld.wolfram.com/SphericalCoordinates.html
33. Williams, E.M., Graham, S.S., Reed, P.A., Rushing, T.S.: Laboratory characterization of cor-tuf concrete with and without steel fibers. Tech. rep, Engineer Research and Development Center Vicksburg MS Geotechnical and Structures Lab (2009)
34. Young, I.T., Gerbrands, J.J., Van Vliet, L.J.: Fundamentals of Image Processing. Delft University of Technology Delft (1998)
35. Zack, G., Rogers, W., Latt, S.: Automatic measurement of sister chromatid exchange frequency. J. Histochem. Cytochem. **25**(7), 741–753 (1977)

Mechanical Characteristics of SFRC Reinforced by New Engineered Steel Fibre

Tomasz Ponikiewski and Jacek Katzer

Abstract The last five decades have seen a large research effort focused on fibre reinforced concrete. Most of the research studies have been devoted to the use of steel fibre and mechanical characteristics of fibre reinforced concrete. Only a limited number of research programmes dealt with properties (both geometrical and mechanical) and development of steel fibre. In this paper a new type of engineered steel fibre will be analysed in comparison to commercially available fibre. Properties of concretes reinforced by a small volume of fibre were of special interest. Conducted tests covered both fresh mix behaviour and strength characteristics of hardened concrete.

1 Introduction

Since the invention of modern steel fibre reinforced concrete (SFRC) in 1874 [7] new types of steel fibre have been engineered every year. Currently, the most popular type of engineered steel fibre is hooked fibre [6], which is produced globally by numerous companies in different sizes and shapes of hook [4]. Other popular types of steel fibre are crimped, coned and fibre with roughened surface [7, 9, 18]. Over the years new types of steel fibre were proposed (e.g. double hooked fibre or twisted fibre) but they did not catch-up construction industry [8]. Research programmes dealing with SFRC usually target high and very high mechanical properties. In such cases, using sophisticated engineered steel fibre which is more expensive than ordinary hooked steel fibre is a sustainable solution. Nevertheless, on a daily basis huge majority of cast SFRC is modified by relatively small volume of fibre and characterized by average compressive strength. The most common fibre dosages range between 25 and 45 kg per cubic meter. In such cases benefits achieved by replacing traditional

T. Ponikiewski
Silesian University of Technology, Gliwice, Poland
e-mail: tomasz.ponikiewski@polsl.pl

J. Katzer (✉)
Koszalin University of Technology, Koszalin, Poland
e-mail: katzer@tu.koszalin.pl

H. Herrmann and J. Schnell (eds.), *Short Fibre Reinforced Cementitious Composites and Ceramics*, Advanced Structured Materials 95,
https://doi.org/10.1007/978-3-030-00868-0_5

steel fibre by new type of engineered steel fibre are usually not noticeable. In authors opinion there is a need for new engineered steel fibre, but the benefits of using it must be clearly "visible" in case of low volumes of added fibre. Properties of the fresh mix are also very important—one prefers to utilize ordinary compaction techniques. Keeping all above facts in mind the research programme was planned and conducted using a new type of fibre.

2 Materials

The composition of a concrete mix was fixed. Portland cement CEM I 32.5 R was used as a binder. The amount of utilized aggregate was equal to 1953 kg per cubic meter. The aggregate was in a form of sand, fine gravel (2–8 mm) and coarse gravel (8–16 mm). All aggregates were of natural origin. The mix was modified by superplasticizer based on polycarboxylate ether characterized by pH = 8 and density of $1160 \pm 30\,kg/m^3$. The superplasticizer was dosed in amount of 2% (by weight of binder). The targeted strength class and consistency was C30/37 and S2 respectively, according to EN 206. The water-cement ration was equal to 0.48. The detailed concrete mix composition is presented in Table 1.

The research programme was conducted using two types of engineered steel fibre. Fibre 50 was chosen as a reference reinforcement. This fibre is hooked, produced on mass scale and commercially available. It is successfully used for both ordinary and self-compacting concretes [10, 11]. Fibre 55 is a new type of fibre with complex hooked and crimped geometry. This fibre is in final phase of experimental development just before entering mass scale production.

Due to its elaborate shape the extra gains in mechanical characteristics of SFRC based on Fibre 55 are expected. The main properties of both fibres are presented in Table 2. The fibre dosage was ranging from 0.0 to 0.57% by volume. Such a small volume of added fibre was chosen deliberately to mirror the most commonly utilized fibre dosage of 25, 35 and $45\,kg/m^3$ in construction industry in Central Europe. The fibre dosage is described in details in Table 3.

Table 1 Concrete mix composition of $1\,m^3$

Ingredient	Amount (kg)
CEM I 32.5 R	350.0
Superplasticizer	7.0
Water	168.0
Sand 0–2 mm	635.0
Gravel 2–8 mm	830.0
Gravel 8–16 mm	488.0
Steel fibre	0.0–45.0

Table 2 Engineered steel fibre

Code name	Length (mm)	Diameter (mm)	Cross section	Shape	
Fibre 50	50	0.65	Circle	Hooked	
Fibre 55	55	0.65	Circle	Hooked & Crimped	

Table 3 Fibre dosage

m_f (kg)	25.0	35.0	45.0
V_f (%)	0.32	0.45	0.57

3 Methods

Concrete mixes were prepared in lab conditions. Cast specimens were kept in moulds for the first 24 h of curing. The moulds were tightly wrapped with polyethylene sheet to maintain high moisture conditions. Subsequently, the specimens were demoulded and placed in a water tank (temp. +20 °C *pm* 0.5 °C) for the next 27 days. The specimens were in a form of cubes (150 mm × 150 mm × 150 mm) and prisms (150 mm × 150 mm × 550 mm). Cubes were used for establishing the density and for the compressive test. Flexural characteristics of the concretes in question was obtained through the tests of the prisms according to EN 14651 (LOP method) [1]. Flexural tensile strength f_{LOP} and residual flexural tensile strengths f_{R1}, f_{R2}, f_{R3} and f_{R4} were acquired during this test. The values of equivalent strengths $f_{eq,2}$ and $f_{eq,3}$ were also established (according to RILEM TC 162-TDF). The needed notches (25 mm deep) located on the bottom surface of the prisms were cut just before the test. A prismatic specimen with a notch during the flexural test is presented in Fig. 1.

Fig. 1 A prismatic specimen with a notch during the flexural test (LOP method)

The flexural test was conducted with constant growth of deflection (δ). The values of crack mouth opening displacement (CMOD) were calculated using the Eq. 1 provided by EN 14651.

$$\delta = 0.85 \times \mathrm{CMOD} + 0.04 \tag{1}$$

Due to unknown behaviour of Fibre 55 during mixing, casting and compaction of fresh concrete mix, its spacing in hardened SFRC was a significant concern. Therefore the additional non-destructive test of SFRC with Fibre 55 was planned. X-ray computed tomography (CT) as a proven method for testing fibre spacing [5, 10, 15–17] was harnessed for this task. Each specimen was subjected to

CT scanning using 64-row scanner and acquisition protocol: 140 kV cathode voltage, 450 mA X-ray tube current and using helical mode. Through the tomography reconstruction procedure volumetric images (512 × 512 × 1500) with 12-bit greyscale were obtained. Used CT apparatus, procedures and software were thoroughly described in previous publications [11]. As the result of the test images with fibre spacing were obtained and its analysis was enabled. Spacing of Fibre 50 was tested with the help of CT multiple times and was described in numerous publications [10, 11] thus unnecessary to be repeated.

4 Results

Consistency of fresh mix was the first tested property. The results of the slump test are presented in Fig. 2. The concrete matrix was characterized by slump h = 50 mm, which is a very minimum value to be classified as consistency class S2. The addition of both types of fibre influenced the consistency. SFRC reinforced by Fibre 50 was characterized by slump of 50, 30 and 20 mm for volume of fibre of 0.32, 0.45 and 0.57% respectively. It can be stated that all SFRC modified by Fibre 50 were fulfilling the requirements of consistency class S1. The slump of all tested concretes with Fibre 55 was smaller than adequate concrete with Fibre 50. SFRC reinforced by Fibre 55 was characterized by slump of 35, 25 and 10 mm for volume of fibre of 0.32, 0.45 and 0.57% respectively. SFRC with 0.32% and 0.45% of Fibre 55 can be classified as consistency class S1. The slump of SFRC with 0.57% of Fibre 55 is out of classification scale defined by EN 206 [2], thus the consistency class can't be assigned.

All cube specimens were used (measured and weighted) to get the apparent density of prepared concretes. The achieved results are presented in Table 4. The apparent density of concrete matrix was equal to 2346 kg/m^3. Adding steel fibre was increasing the apparent density which reached the maximum value of 2378 kg/m^3 in case of concrete with 0.57% of Fibre 50. This is a difference of 1.4% and should be considered as irrelevant for the analysis of other properties.

The lowest compressive strength tested on cubes was registered for concrete matrix. The compressive strength was getting higher alongside added fibre.

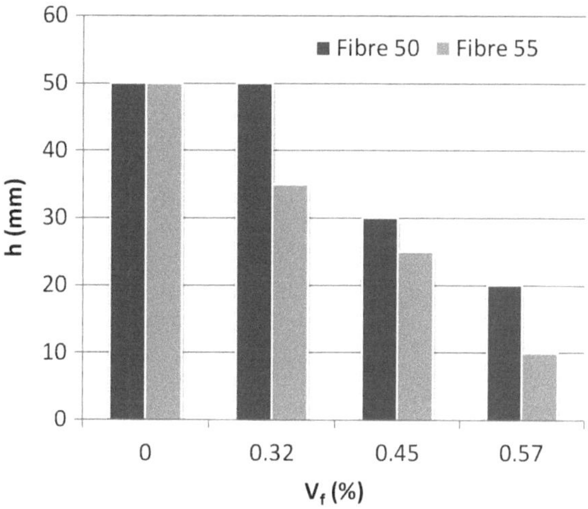

Fig. 2 Consistency

Table 4 Compressive strength and density

	Matrix	Fibre 50			Fibre 55		
V_f (%)	0.00	0.32	0.45	0.57	0.32	0.45	0.57
$f_{c,cube}$ (MPa)	56.6	60.1	62.8	63.6	61.6	61.5	66.0
ρ (kg/m^3)	2346	2352	2365	2378	2339	2359	2375

The highest values were achieved by concretes reinforced by 0.57% of fibre. The largest difference of strength between concrete matrix and concrete with maximum added volume of Fibre 55 was equal to 16.6%. Registered strength values allowed to assign concrete strength classes according to EN 206 [2]. Concrete matrix can be assigned to concrete strength class C40/50 and concrete with 0.57% of Fibre 55 can be assigned to concrete strength class C50/60. All other tested concretes are assigned to concrete strength class C45/55 (Fig. 3).

The values of f_{LOP} and equivalent flexural strengths $f_{eq,2}$ and $f_{eq,3}$ are presented in Figs. 4 and 5 respectively. The addition of fibre significantly influences both flexural strengths. In case of strength $f_{eq,2}$ the addition of 0.57% of Fibre 50 improves the achieved value by 480% and the addition of Fibre 55 by 529% in comparison to concrete matrix. The largest differences in efficiency of Fibre 50 and Fibre 55 can be observed for the smallest volume of added fibre. In case of $V_f = 0.32\%$ of Fibre 50, $f_{eq,2}$ increased by 14% and in case of Fibre 55 by 260%.

It is also worth noticing, that equivalent flexural tensile strength $f_{eq,2}$ achieved by concrete with the addition of 0.45% of Fibre 50 is only 15% higher than strength of concrete with the addition of 0.32% of Fibre 55. Similar relations characterize values of $f_{eq,3}$. Fibre 55 proves to be more efficient reinforcement than Fibre 50 for all fibre volumes. There is also a huge visible difference in strength values for the

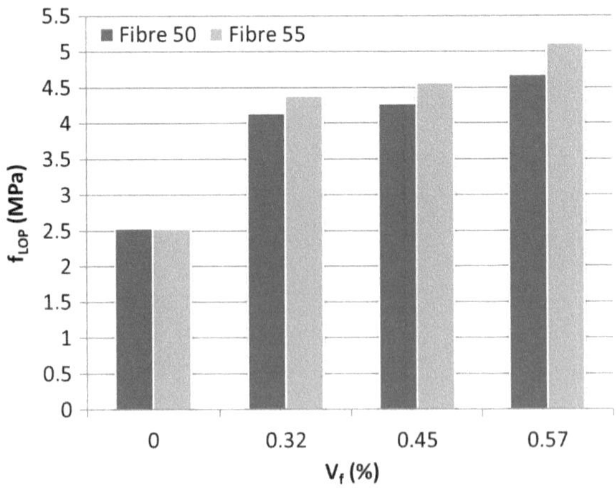

Fig. 3 Flexural tensile strength

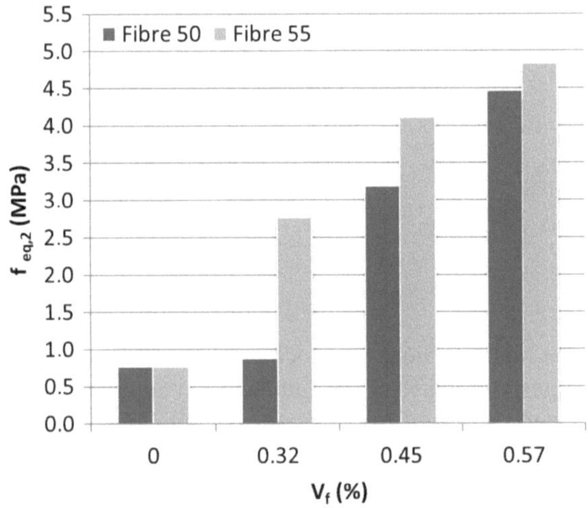

Fig. 4 Equivalent flexural tensile strength $f_{eq,2}$

smallest amount of fibre. Concrete with Fibre 55 is basically twice "stronger" then concrete with Fibre 50. Results of flexural tensile strength f_{LOP} and all four residual strengths are presented in Table 5. Residual strengths corresponding to different values of the CMOD are fairly difficult to incorporate to SFRC designing methods [12]. It is commonly assumed that residual strength f_{R1} and f_{R3} are associated with serviceability limit states (SLS) and ultimate limit states (ULS) respectively. According to fib Model Code fibre concrete behaviour at ULS is related to the behaviour at SLS through the ratioof f_{R3}/f_{R1}. This ratio is used for classification of SFRC. Post-cracking residual strength is described by two parameters: f_{R1}—representing the strength interval and a letter a, b, c or d representing the f_{R3}/f_{R1} ratio. Such a classification mirrors four most common cases of SFRC softening and hardening. The fib Model Code allows to assess possible traditional reinforcement

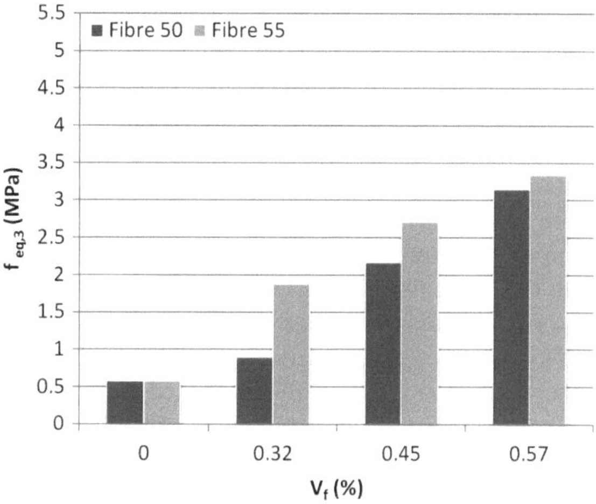

Fig. 5 Equivalent flexural tensile strength $f_{eq,3}$

Table 5 Flexural tensile strength and residual strengths

	V_f %	f_{LOP} MPa	f_{R1} MPa	f_{R2} MPa	f_{R3} MPa	f_{R4} MPa	f_{R3}/f_{R1} (> 0.5)	f_{R1}/f_{LOP} (> 0.4)	Strength class	Reinforcement substitution
Matrix	0.00	2.54	n/a	n/a	n/a	n/a	–	–	–	–
Fibre 50	0.32	4.14	1.40	1.02	0.74	0.63	0.528	0.338	1a	Disabled
Fibre 50	0.45	4.28	3.25	2.11	1.57	0.96	0.483	0.759	–	Disabled
Fibre 50	0.57	4.68	4.35	3.12	2.22	1.50	0.510	0.929	4a	Enabled
Fibre 55	0.32	4.39	2.63	1.88	1.25	3.45	0.475	0.599	–	Disabled
Fibre 55	0.45	4.57	3.81	2.94	1.64	1.11	0.430	0.834	–	Disabled
Fibre 55	0.57	5.12	4.74	3.54	2.39	1.73	0.504	0.926	4a	Enabled

(re-bars and stirrups) substitution by fibre. The assessment is based on two ratios f_{R1}/f_{LOP} and f_{R3}/f_{R1}. In Table 5 both ratios are presented and compared with minimum requirements. Only three out of six tested SFRC fulfilled the minimum requirements to be assigned a strength class. The crucial factor for this task was the value of ratio f_{R3}/f_{R1}. Only two out of six tested SFRC are enabled to substitute traditional reinforcement (both with $V_f = 0.57\%$).

5 Discussion

New hooked and crimped types of engineered steel fibre proved to be more efficient than traditional hooked fibre. All tested and calculated strength parameters were higher in case of concretes with Fibre 55. The largest differences were noted for smallest volumes of added fibre. This phenomenon is very important for construction industry due to a common practice of using small volumes of fibre. On the other hand only concretes with $V_f = 0.57\%$ were characterized by mechanical properties enabling strength class assignment and reinforcement substitution. Thus, only these SFRC would have a structural significance. Both SFRC with $V_f = 0.57\%$ were described by strength class 4a. It is again proven that the minimum volume of added fibre should be over 0.5% [3]. All attempts to minimalize the volume of fibre are not sustainable form the structural point of view. The only disadvantage associated with Fibre 55 is consistency. Mixes with Fibre 55 are characterized by lower slump in comparison to mixes with Fibre 50. The largest difference was observed for SFRC with 0.32% of fibre. Nevertheless, the differences range from 10 to 15 mm and in majority of cases the same consistency class is maintained.

Non-destructive CT test (see Fig. 6) allowed assessment of spacing of Fibre 55. In general Fibre 55 is uniformly distributed in the concrete volume. Spacing of hooked and crimped fibre is similar to spacing of hooked 50 (which was described in previous publications [10, 11]). Therefore all modelling [13, 14] and design approaches which were useful in case of hooked fibre should be valid for Fibre 55.

Fig. 6 CT images of SFRC specimens' cross-sections with Fibre 55 (from left $V_f = 0.32\%, 0.45\%, 0.57\%$)

6 Conclusions

The conducted research programme allow to draw following conclusions:

- The new type of fibre with complex hooked and crimped geometry is more efficient reinforcement than traditional hooked fibre,
- Hooked and crimped fibre is especially efficient reinforcement when added in small volumes in comparison to hooked fibre,
- Hooked and crimped fibre influences the consistency of fresh concrete mix in a more significant way than hooked fibre,
- Research programme focused on concretes with larger volumes of hooked and crimped fibre should be conducted,
- Properties of SFRC with hooked and crimped fibre should be compared with concretes with crimped fibre and other hooked fibre.

References

1. En 14651 test method for metallic fibered concrete—measuring the flexural tensile strength (limit of proportionality (lop), residual)
2. En 206 concrete - specification, performance, production and conformity
3. Domski, J.: A blurred border between ordinary concrete and SFRC. Constr. Build. Mater. **112**, 247–252 (2016)
4. Domski, J., Katzer, J., Zakrzewski, M., Ponikiewski, T.: Comparison of the mechanical characteristicsof engineered and waste steel fiber used as reinforcement for concrete. J. Clean. Prod. **158**, 18–28 (2017)
5. Herrmann, H., Pastorelli, E., Kallonen, A., Suuronen, J.P.: Methods for fibre orientation analysis of x-ray tomography images of steel fibre reinforced concrete (SFRC). J. Mater. Sci. **51**(8), 3772–3783 (2016). https://doi.org/10.1007/s10853-015-9695-4
6. Katzer, J., Domski, J.: Quality and mechanical properties of engineered steel fibres used as reinforcement for concrete. Constr. Build. Mater. **34**, 243–248 (2012)
7. Maidl, B.R.: Steel Fibre Reinforced Concrete. Ernst & Sohn, Berlin, Germany (1995)
8. Naaman, A.E.: Engineered steel fibres with optimal properties for reinforcement of cement composites. J. Concr. Adv. Technol. **1**(3), 241–252 (2003)
9. Nawy, E.G.: Fundamentals of High Strength High Performance Concrete. Longman, England (1996)
10. Ponikiewski, T., Katzer, J., Bugdol, M., Rudzki, M.: Determination of 3D porosity in steel fibre reinforced SCC beams using x-ray computed tomography. Constr. Build. Mater. **68**, 333–340 (2014). https://doi.org/10.1016/j.conbuildmat.2014.06.064
11. Ponikiewski, T., Katzer, J., Bugdol, M., Rudzki, M.: Steel fibre spacing in self-compacting concrete precast walls by x-ray computed tomography. Mater. Struct. **48**(12), 3863–3874 (2015). https://doi.org/10.1617/s11527-014-0444-y
12. di Prisco, M., Plizzari, G., Vandewalle, L.: Fibre reinforced concrete: new design perspectives. Mater. Struct. **42**(9), 1261–1281 (2009). https://doi.org/10.1617/s11527-009-9529-4
13. Seitl, S., Kersner, Z., Bilek, V., Knesl, Z.: Fatigue parameters of cement based composites with various types of fiber. Key Eng. Mater. **417–418**, 129–132 (2012)
14. Seitl, S., Simonova, H., Kersner, Z., Fernandez-Canteli, A.: Evaluation of concrete fatigue measurements using standard and non-linear regression model. Appl. Mech. Mater. **10**, 121–126 (2011)

15. Suuronen, J.P., Kallonen, A., Eik, M., Herrmann, H., Serimaa, R.: Determination of fiber orientation in steel fiber reinforced concrete by x-ray microtomography. In: Proceedings of Physics Days 2012, p. 188. The Finnish Physical Society (2012)
16. Suuronen, J.P., Kallonen, A., Eik, M., Puttonen, J., Serimaa, R., Herrmann, H.: Analysis of short fibres orientation in steel fibre reinforced concrete (SFRC) using x-ray tomography. J. Mater. Sci. **48**(3), 1358–1367 (2013). https://doi.org/10.1007/s10853-012-6882-4
17. Vicente, M.A., González, D.C., Mínguez, J.: Determination of dominant fibre orientations in fibre-reinforced high-strength concrete elements based on computed tomography scans. Nondestr. Test. Eval. **29**(2), 164–182 (2014). https://doi.org/10.1080/10589759.2014.914204
18. Zollo, R.F.: Fiber-reinforced concrete: an overview after 30 years of development. Cement Concr. Compos. **19**(2), 107–122 (1997). https://doi.org/10.1016/S0958-9465(96)00046-7

Short Composite Fibres for Concrete Disperse Reinforcement

Arturs Lukasenoks, Andrejs Krasnikovs, Arturs Macanovskis, Olga Kononova and Videvuds Lapsa

Abstract Short composite fibres are a relatively new product for concrete disperse reinforcement. In this experimental research, 14 different composite fibres were developed and single fibre pull-out micromechanics was investigated. Three main groups of composite fibre were—composite glass fibres (GF), composite carbon fibres (CF) and hybrid fibres (HF). Composite fibre manufacturing consisted of glass, carbon or combined fibre filament preparation, impregnation with epoxy resin, epoxy curing, quality control and cutting in short discrete macro- fibres. All three composite fibre groups were manufactured with straight, uneven and undulated geometries. Fibre surface finish was smooth and rough. Uneven fibre geometry was achieved by not aligning all fibre filaments in fibre tow. Undulated geometry was a result of interlaced fibres. The rough fibre outer surface finish was achieved by adding an extra layer of epoxy resin containing fine quartz grains. All macro-fibres were cut in 50 mm length. Single fibre pull-out samples with a pre-defined crack between two concrete parts were prepared to investigate fibre pull-out behaviour. Fibre pull-out laws were obtained and analysed. Composite fibre improvement geometry and surface with roughening outer surface made a huge impact on fibre pull-out resistance.

A. Lukasenoks (✉) · A. Krasnikovs · A. Macanovskis · O. Kononova
Concrete Mechanics laboratory, Institute of Mechanics,
Riga Technical University, Riga, Latvia
e-mail: Arturs.Lukasenoks@rtu.lv

A. Krasnikovs
e-mail: Andrejs.Krasnikovs@rtu.lv

A. Macanovskis
e-mail: Arturs.Macanovskis@rtu.lv

O. Kononova
e-mail: Olga.Kononova@rtu.lv

V. Lapsa
Faculty of Civil Engineering, Institute of Building Production,
Riga Technical University, Riga, Latvia
e-mail: Videvuds.Lapsa@rtu.lv

H. Herrmann and J. Schnell (eds.), *Short Fibre Reinforced Cementitious Composites and Ceramics*, Advanced Structured Materials 95,
https://doi.org/10.1007/978-3-030-00868-0_6

1 Introduction

Development of composite fibres is very important due to many problems related to polymer, glass and carbon fibre usage in concrete for disperse reinforcement. There is a necessity to use large volume fractions 2–2.5% of polymer fibre to achieve a tangible increase in load bearing performance of low strength concretes [1]. Large amounts of fibres significantly worsen concrete consistency, a new approach for concrete design with target consistency is necessary. Fibres with higher strength and elasticity modulus are necessary for higher strength concretes. Due to fibre small diameter glass, basalt or carbon fibres have significant problems with introducing in concrete. When fibres are introduced into the concrete mix, fibre rolls and clews are formed—problems with fibre homogeneity in the concrete occurs. Each fibre roll or clew with air voids is another defect in the concrete which reduces strength. Difficulties with fibre introduction become more pronounced with fibre volume fraction increase. Mentioned fibres can be introduced in concrete mixes produced in high-speed mixers, unfortunately, high mixing energy and shear forces brakes and mills fibres in shorter parts [2]. Small fibres can work bridging only small crack openings.

Glass, basalt and carbon fibres are used more widely due to advanced manufacturing technologies and reduced cost. Several attempts to introduce glass, basalt and carbon fibres in the construction industry were found—long carbon fibres and commercial product Minibar [3, 4]. Basalt Minibar fibres are recommended to use in high volume fractions (1.5–3%) to achieve tangible results. Long carbon fibre composite fibres were used for increasing concrete plate's impact load resistance [4, 5].

In the present investigation, short composite macro-fibres were manufactured as unidirectional carbon and glass composite rods. Rods were cut in discrete fibres with L/d ratio from 18.1 to 56.1. A varying number of filaments in macro-fibre (changing filament volume fraction V_f in composite fibre) as well as using different material filaments in one macro-fibre it is possible to obtain reinforcing fibres with different strengths and elastic properties. Composite fibres can be developed for very specific use; composite fibres with a smooth surface for high strength concretes to increase ductility, fibres with a rough surface or uneven geometry results in good anchorage for low and average strength concrete reinforcement [6]. Higher fibre anchorage in concrete can be also achieved by making fibre outer surface rough [7].

2 Materials and Methods

Overview of Experimental program for composite fibres and different concrete matrices is shown in Table 1.

Table 1 Concrete mixture proportions

Fibre type	Fibre denomination	Concrete matrix			Depth, angle configuration
		M1	M2	M3	
Composite glass fibres	GF1	-	-	x	25 mm depth, 0-degree angle to pull-out force
	GF2	x	x	x	
	GF3	-	-	x	
	GF4	-	-	x	
	GF5	-	-	x	
Composite carbon fibres	CF1	x	x	x	
	CF1-A	x	x	x	
	CF1-B	x	x	x	
	CF2	x	x	x	
	CF2-A	x	x	x	
	CF2-B	x	x	x	
	CF3	-	x	x	
Hybrid fibres	GC1	-	-	x	
	CG1	-	-	x	

3 Concrete Materials and Mix Design

Portland cement Aalborg White CEM I 52.5 R is used as binder in experimental mixes, naturally fractioned and washed quartz sand 0–1 mm as main aggregate. Quartz powder and silica fume was used as micro filler. Poly-carboxylate based high range water reducing admixture was used to control mix workability. Tree types of concrete mixtures were designed (mixture proportions are presented in Table 2). First type concrete mixture is high strength concrete (M1) with cement content 800 kg/m^3), with silica fume, water to cement w/c ratio 0.25. The second type concrete mixture is normal strength concrete (M2) with cement content 550 kg/m^3, having silica fume and with w/c ratio equal to 0.55. The third type concrete mixture was low strength (M3) with cement content 400 kg/m^3, without silica fume, and having water to cement ratio w/c equal to 0.7. The amount of micro-filler was adjusted in order to achieve paste content 550 l $\pm$ 7% in all cases.

All three concrete types were designed to achieve target compressive strength and have good workability and stability in fresh stage. Workability was defined to have slump flow classes measured in accordance with EN 12350-8: 580 mm.

Table 2 Concrete mixture proportions

Material	Density, t/m^3	M1	M2	M3
Cement CEM I 52,5 R (*Aalborg White*)	3,13	800	550	400
Water	1	200	300	300
Sand 0–1 mm (*SaulkalneS*)	2,65	1100	1200	1400
Microsilica (*Elkem 920D*)	2,22	133,3	50	0
Quartz powder 0–120 mk (*Anyksciai*)	2,65	66,5	250	250
HRWR (*Sika D400*)	1,07	25	6,5	3
w/c ratio		*0,25*	*0,55*	*0,75*
paste volume, l		*605,5*	*611,4*	*516*

4 Fibres

Composite fibres were developed using filament tows—glass, carbon and combined both. Different types of fibres were manufactured—with smooth, rough and undulated surface.

There were used several manufacturing techniques to achieve different fibre geometry and surface, but general approach was—fixing tow with fibre filaments in setup, impregnation with epoxy resin, curing and cutting into 50 mm long discrete fibres. Summary of fibre types and properties can be found in Table 3.

Table 3 Fibre properties

Fibre type	Fibre denomination	Fibre geometry, surface	Diameter, mm	L/d ratio	Reinforcement ratio	Specific weight, kg/m^3
Composite glass fibres	GF1	Uneven, smooth	2,18	22,9	0,534	1195
	GF2	Straight, smooth	1,52	32,9	0,751	1746
	GF3	Straight, rough	2,04	24,5	0,482	1507
	GF4	Undulated, smooth	1,88	26,6	0,657	1310
	GF5	Undulated, smooth	1,85	27,0	0,696	1270
Composite carbon fibres	CF1	Straight, smooth	1,77	28,3	0,578	1129
	CF1-A	Straight, smooth	1,30	38,6	0,635	1291
	CF1-B	Straight, smooth	0,89	56,1	0,551	1569
	CF2	Straight, rough	2,10	23,8	0,335	1397
	CF2-A	Straight, rough	1,79	27,9	0,321	1330
	CF2-B	Straight, rough	1,57	31,8	0,235	1183
	CF3	Undulated, smooth	1,82	27,5	0,566	1101
Hybrid fibres	GC1	Undulated, smooth	2,60	19,2	0,755	997
	CG1	Undulated, smooth	2,76	18,1	0,748	894

5 Composite Glass Fibres

Five types of fibres (GF1–GF5) were developed using glass filaments. Manufacturing of GF1 fibres is simplest—fibres were tied in the upper part of the fibre tow and fibre filaments were free falling during the epoxy resin impregnation. Several fibre filaments were dropping out from perfect order and uneven fibre geometry was created. Manufacturing of GF2 was more complicated—each fibre filament was aligned to fit perfectly in the fibre tow, so straight fibre was created. Slight fibre pre-stress was realized with springs, before epoxy resin was applied. Fibre pre-stress was made to ensure fibre straight geometry and smooth surface. Fibres GF1 and GF2 are shown in Fig. 1.

Manufacturing of GF3 is the same as GF2, but the main difference is final epoxy resin layer creating fibre surface finish rough. After fibre filament impregnation with epoxy resin another part of epoxy was prepared with fine quartz grains and applied. Final layer applied on the fibre surface using brush. Manufacturing of GF4 and GF5 is the same, main difference is preparing fibre tows. GF4 and GF5 fibres are made from interlaced fibre filaments. GF4 fibre was made of interlaced fibre tows creating 14 braids and GF5 had interlacing much looser creating 7 braids (Fig. 2).

6 Composite Carbon Fibres

Manufacturing of carbon fibres CF1, CF1-A and CF1-B is the same—fibre filament tow was tied in manufacturing setup and impregnated with epoxy resin. CF1 has 24 k carbon fibre filaments, CF1-A has 16 k and CF1-B has 8 k fibre filaments.

Fig. 1 Composite glass fibres—GF1 on the left and GF2 on the right

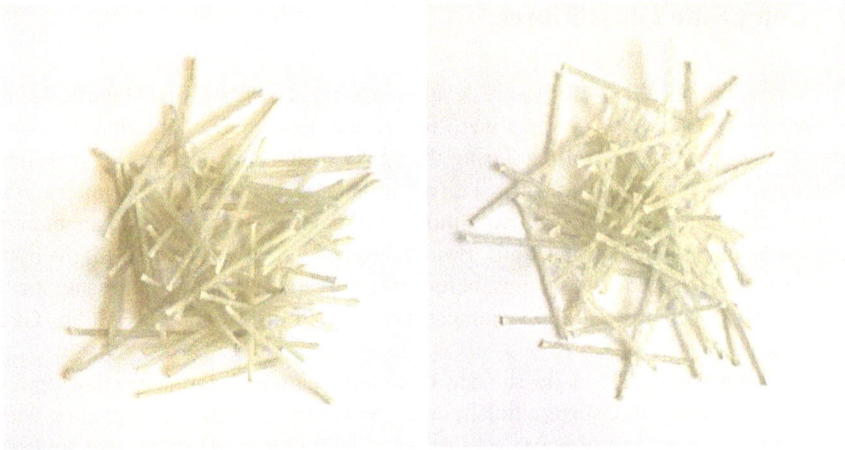

Fig. 2 Composite glass fibres—GF3 on the left and GF4 on the right

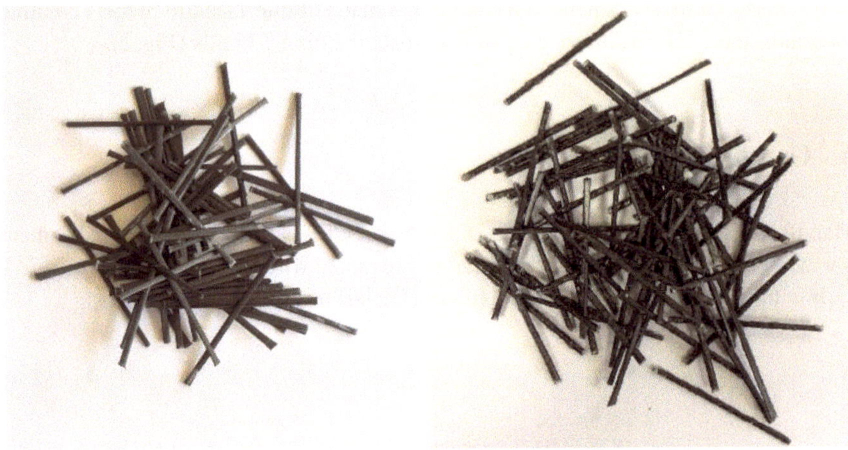

Fig. 3 Composite carbon fibres—CF1 on the left and CF2 on the right

Group of CF2 fibres (CF2, CF2-A, CF2-B) was manufactured in the same way as the group of CF1 fibres, but the main difference was final layer of epoxy. Similarly, to manufacturing of GF3, epoxy resin was prepared with fine quartz grains and applied on fibres CF2, CF2-A and CF2-B as final coating on the fibre surface (Fig. 3).

Manufacturing of CF3 consists of 24 k filament fibre tow preparation, dividing in three equal parts and interlacing creating 7 braids on one 50 mm long fibre. Prepared fibre tow with interlaced braids was tied in manufacturing setup, epoxy resin was applied and cured. Brush was used to apply epoxy resin due to fibre fluted surface (Fig. 4).

Fig. 4 Composite carbon fibres CF3

7 Hybrid Fibres

Manufacturing of hybrid fibres CG1 and GC1 consisted of combining two types of fibre filaments (glass and carbon) in one composite fibre. These fibres were manufactured by making a braid from one type of fibres and during the process another fibre is weaved into with certain step to form a hybrid fibre. The first objective for combining fibre types was to strengthen glass fibres with carbon fibre filaments. Second objective was to create uneven un undulating surface to improve hybrid fibre and concrete matrix bond and grip during pull-out process (Fig. 5).

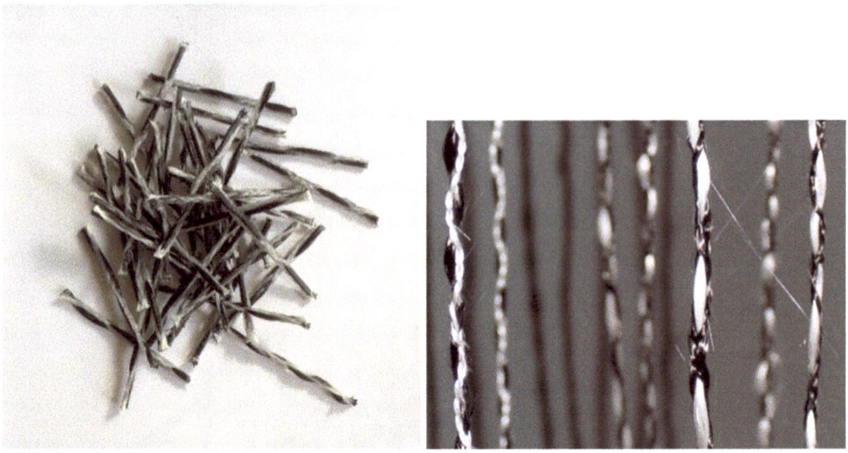

Fig. 5 Hybrid fibres—CG1 on the left, both CG1 and GC1 fibres hanging before epoxy resin application on the right

8 Samples

Single fibre pull-out experiments were carried out to investigate fibre resistance to pull-out from concrete matrix. There different concretes were developed to investigate fibre pull-out behaviour in concretes with various strengths. Fibres were embedded—25 mm in depth.

All experimental tests are performed at 28-day age of concrete cured in normal conditions RH > 95%, 20 °C. Concrete compressive strength was tested. Single fibre pull-out tests were carried out using testing machine Zwick Z150. Pull-out displacement was measured using non-contact measuring device—video extensometer *Messphysik*. Loading rate—5 mm/min was used for all specimen pull-out tests. Test specimens are made with pre-defined crack, configured in depth and angle. Pre-defined crack ensures that only fibre bridges forces between two concrete parts.

9 Results and Discussion

Average fresh concrete slump flow and compressive strength results for concrete matrix types M1, M2 and M3 are presented in Table 4.

Average pull-out curves for composite fibres are calculated and presented in Figs. 6, 7, 8 and 9. Stress and deformation is plotted on the charts to show force and fibre cross section area influence on the pull-out process. Composite fibres CF1, CF1-A, CF1-B, CF2, CF2-A, CF2-B, CF3 and GF2 are tested in all three concrete matrices M1, M2 and M3. Fibres GF1, GF3, GF4, GF5, CG1 and GC1 are tested in matrix M3.

Table 4 Concrete matrix properties

Concrete property	M1	M2	M3
Fresh concrete consistency—slump flow, mm	768	743	580
Concrete compressive strength 28-day age, MPa	124.44	71.9	34.09

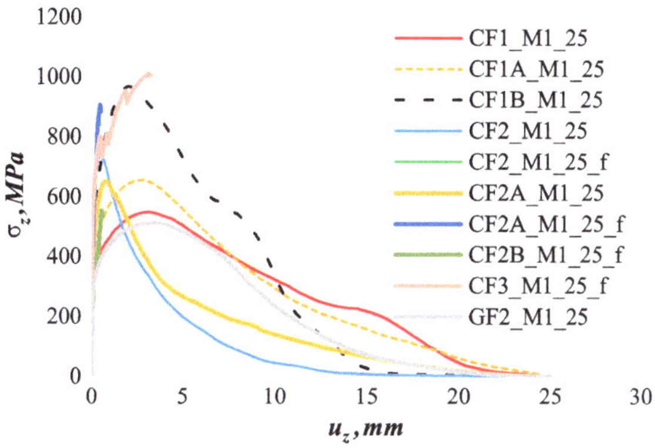

Fig. 6 Average single fibre pull-out behaviour curves (stress—deformation) for group of carbon fibres (CF1–CF3) and glass fibre (GF2), embedded in 25 mm depth. Concrete matrix: M1

Smooth composite carbon fibres have different stress levels when pulling out from all three concrete matrices. Pull-out stress curve is lowest for composite fibre CF1 and curve peak increasing 1.3 times for CF1-A and 1.95 times CF2-B.

Results from the Fig. 6 show that fibres CF2-A in M1 with increased roughness on the outer surface can achieve the same stress level as CF1-A and some of the tested fibres even failed because of achieving higher grip with concrete matrix. In the same way, CF2-A in M1 has faster pull-out resistance drop compared to CF1-A due to external layer peeling off. Force level is still high, but due to reduction in diameter calculation takes in account initial diameter. Performance of the same fibres are dramatically different in concrete matrix M2 due to high relation between smooth fibre pull-out resistance and concrete strength. CF2-A fibre has more than twice higher stress level during the initial pull-out and dropping at 5 mm to nearly the same as CF1-A.

GF2 fibre in M3 is the only glass fibre with high fibre delamination stress. This fibre has the smoothest surface compared to other glass fibre surfaces. GF1, GF4 and GF5 fibre fails at different stress levels (250–312 MPa), these fibres have one common feature—their geometry is complicated which results in high fibre anchorage in concrete matrix. GF1 has uneven shape due to manufacturing process and GF4, GF5 have undulated shape due to interlacing fibre braids.

From the Figs. 8 and 9 it can be seen that higher resistance can be achieved when hybrid fibre manufactured with central glass fibre filaments and circumjacent carbon fibres—CG1. Hybrid fibre CG1 has 15% higher peak resistance compared to GC1.

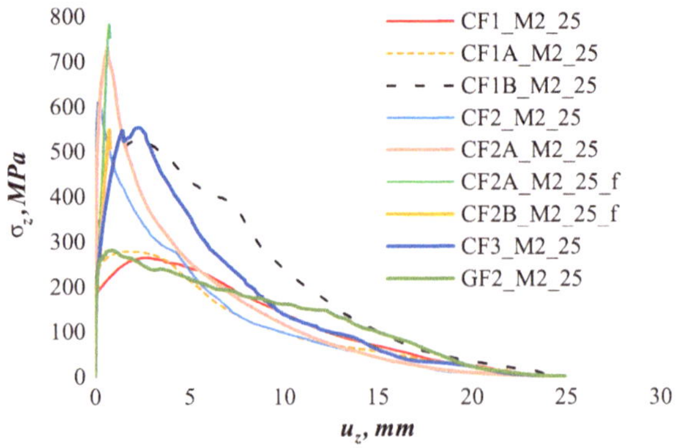

Fig. 7 Average single fibre pull-out behaviour curves (stress—deformation) for group of carbon fibres (CF1–CF3) and glass fibre (GF2), embedded in 25 mm depth. Concrete matrix: M2

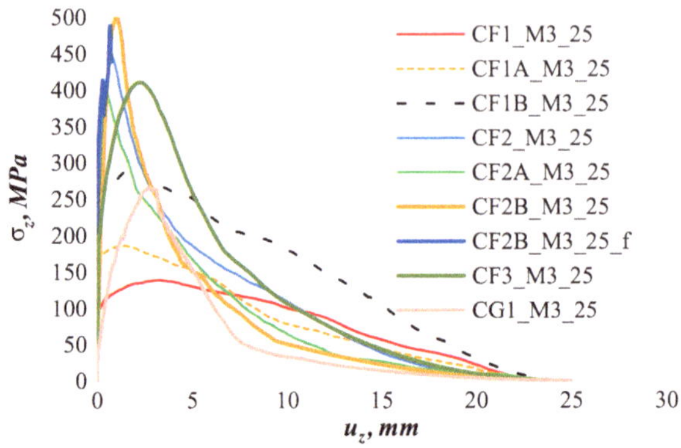

Fig. 8 Average single fibre pull-out behaviour curves (stress—deformation) for group of carbon fibres (CF1–CF3) and hybrid fibre (GC1), embedded in 25 mm depth. Concrete matrix: M3

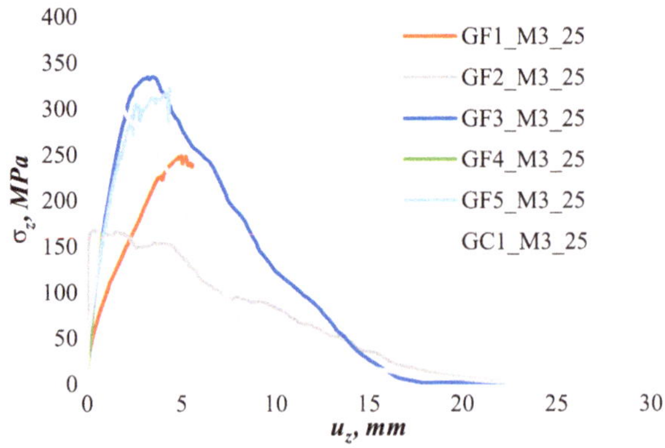

Fig. 9 Average single fibre pull-out behaviour curves (stress—deformation) for group of glass fibres (GF1–GF5) and hybrid fibre (CG1), embedded in 25 mm depth. Concrete matrix: M3

10 Conclusion

Mechanical load bearing capacity of the novel type of disperse concrete reinforcement—short polymer composite fibres was investigated experimentally performing single fibre pull-out tests. Results demonstrate that particular composite fibre pulling out of the concrete force is dependent on concrete matrix compressive strength, concrete matrix shrinkage during maturing, different fibres strength, fibres geometry and each fibre outer interlayer mechanical properties. Concrete compressive strength increase is leading to higher values of pull-out force. Adding fibre outer surface roughness is increasing pull-out force and for some type of fibres are leading to fibre rupture. Depending on concrete compressive strength and shrinkage, optimal composite fibre structure can be obtained (length, diameter and outer interlayer thickness).

References

1. Song, P., Hwang, S., Sheu, B.: Strength properties of nylon and polypropylene fibre reinforced concrete. Cem. Concr. Res. **35**(8), 2005 (2005)
2. Sahmenko, G., Krasnikovs, A., Lukasenoks, A., Eiduks, M.: Ultra high performance concrete reinforced with short steel and carbon fibers. In: Environment. Technology. Resources: Proceedings of the 10th International Scientific and Practical Conference, Latvia, Rezekne, pp. 193–199 (2015)
3. Patnaik, A., Miller, L., Adhikari, S., Standal, P.C.: Basalt FRP minibar reinforced concrete. In: Fibre Concrete 2013, Prague, Czech Republic (2013)
4. Patnaik, A., Miller, L., Standal, P.C.: Fiber reinforced concrete made from basalt FRP minbar. In: Proceedings of the 1st Concrete Innovation Conference (CIC), Oslo, Norway (2014)
5. Tabatabaei, Z., Volz, J., Keener, D., Gliha, B.: Comparative impact behavior of four long carbon fiber reinforced concretes. Mater. Des. **55**, 212–223 (2014)
6. Lukašenoks, A., Mačanovskis, A., Krasņikovs, A., Lapsa, V.: Composite fiber pull-out in concretes with various strengths. In: 15th International Scientific Conference Engineering for Rural Development: Proceedings, Latvia, Jelgava, vol. 15, pp. 1417–1423 (2016)
7. Krasnikovs, A., Lapsa, V., Lukasenoks, A.: Composite material's fibre and its production process. Latv. Pat. P, 16–101 (2016)

Influence of the Flow of Self-Compacting Steel Fiber Reinforced Concrete on the Fiber Orientations, a Report on Work in Progress

Heiko Herrmann, Oksana Goidyk and Andres Braunbrück

Abstract This paper presents a report about work in progress of research on the influence of the flow of SCFRC on the fiber orientations. Mechanical properties of the short steel fiber reinforced cementitious materials mostly depend on the fiber orientation and spatial dispersion. Many studies have shown that it is possible to achieve the desired fiber orientation by optimizing the parameters of rheological properties or the casting process. In order to improve the key mechanical properties, multiple statistical experiments with various factors are needed. This paper analyzes the influence of casting velocity and formwork surface quality on the fiber distribution and orientation. A suitable technique for our method was to replace Steel Fiber Reinforced Self-Compacting Concrete (SFRSCC) by a transparent polymer with similar rheological properties as SFRSCC. Preliminary analysis of the experimental results shows that the fibers tend to orient mostly perpendicular to the flow direction and turn their orientation longitudinally near the walls. Experiments showed that the fiber spatial distribution was affected by the casting velocity. Faster casting velocities provided more preferable homogeneous distribution. Moreover, the roughness of the bottom of the formwork demonstrated some influence on the fiber orientations but no significant impact on the spatial dispersion. In addition, we used the image analysis method to estimate fiber orientation and distribution.

1 Introduction

Steel fiber reinforced self-compacting concrete (SFRSCC) is increasingly used in today's building industry because of its alleged advantages over ordinary concrete, however, the use in load-bearing structures requires additional reinforcement. Self-compacting concrete (SCC) itself simplifies the casting process significantly due to

H. Herrmann (✉) · O. Goidyk · A. Braunbrück
Department of Cybernetics, Tallinn University of Technology, Tallinn, Estonia
e-mail: hh@cens.ioc.ee

A. Braunbrück
e-mail: andres@cens.ioc.ee

H. Herrmann and J. Schnell (eds.), *Short Fibre Reinforced Cementitious Composites and Ceramics*, Advanced Structured Materials 95,
https://doi.org/10.1007/978-3-030-00868-0_7

the compacting capacity to entirely fill the formwork under own weight. The essential advantage of using SCC is that it produces materials free of air voids and honeycombs without using additional vibration [1, 2]. According to many studies, the use of short steel fibers for reinforcing concrete has some advantages in the properties of the material, for example, reduced brittleness, improved ductility, flexural and shear strength of the material. Moreover, the fibers reduce cracking, drying shrinkage and permeability of concrete and bridge cracks during loading and transfer of the load [3–8].

Many studies have attempted to estimate the significance of fibers inside the concrete matrix [7–10]. To evaluate this significance properly, it is essential to take into account many factors. The properties of the short steel fiber concrete largely depend on the full cycle of stages of producing SFRSCC from the mixing to the hardening state. Firstly, the recipe of a mixture, the fiber content, fiber aspect ratio, type, geometry and sizes of fibers are crucial. Secondly, it is required to consider the rheological properties of concrete in a fresh state before casting (flowability, viscosity, compacting ability, etc.). Thirdly, the casting and flow parameters, velocity, possible vibrations and fiber orientations inside the concrete matrix after casting have major influence on the properties of concrete [1, 9, 11–14]. In its turn, the fiber orientations are also largely influenced by the flow of concrete, which depends on the formwork geometry and formwork surface quality. Supposedly, a rough and sticky formwork surface produces a different fiber distribution and orientation compared to a smooth and non-sticky one [12, 15].

To obtain the desired fiber orientation that provides the improved mechanical properties of material, the optimization of casting process is needed. For this purpose, to detect the defects of the internal structure of the materials and to evaluate the fiber orientation and distribution inside the cementitious composites, large variety of non-destructive and destructive methods have been proposed. The current research mostly focuses on the non-destructive methods for estimation of fiber orientation and distribution due to some significant benefits. The crucial advantages of non-destructive methods in comparison with destructive are accuracy, reliability, efficiency, cost saving and safety. The last benefit is extremely important, because of most tests are completely harmless to people and all testing methods leave examined samples totally undamaged. The most popular and frequently applicable in practice in the testing of cementitious materials are x-ray Computed Tomography scanning [9, 16–20], image analysis [10], conductometric (AC-IS) [21], electromagnetic [22, 23], ultrasonic [24] and acoustic [25–27] methods.

Together with non-destructive fiber orientation control methods, numerical simulations of fiber concrete and modeling the cracking behavior are widely used in cementitious materials research.

Computational fluid dynamics (CFD) simulations are starting to become used in the planning of SCC castings to investigate if the SCC flows around reinforcement bars and can fill the complete formwork. CFD simulations need to be calibrated to benchmark cases, to make sure the numerical scheme works. For SCC those benchmark cases have been proposed in [28]. However, only the final result, not the filling velocity was compared. For SFRSCC the situation is much more complex and factors that influence the fiber orientation and spatial distribution need to be determined in order to propose relevant benchmark cases. Promising simulations have been performed by Svec et al. [12], using a Lattice-Boltzmann approach and immersed particles, and by Herrmann and Lees [15], using a Finite Volume scheme for a Bingham-plastic and a tensorial equation for the fiber orientation distribution. However, before these methods can be used on an everyday basis in production it must be certain that the influential parameters in the casting process are identified. The main purpose of the current research is to identify those parameters by analyzing experimentally the fiber orientation under different casting factors and establish the correlation between the fiber orientation, formwork surface quality and casting velocity; moreover, to compare different casting methods and conditions.

Therefore, in this study numerical and experimental "simulations" of the fibers inside of concrete mass are compared to real castings. The essence of this approach is that the opaque concrete is replaced by a transparent and viscous polymer solution that is similar to the rheological properties and parameters of the fresh self-compacting concrete. For an initial comparison a small concrete plate was produced and subjected to x-ray Computed Tomography to evaluate the fiber distribution and orientation. The results obtained from the casting experiments with the transparent polymer matrix in this research and those from the experiments with a concrete sample will be compared.

2 Materials and Methods

In this section we will describe two different approaches to estimate the fiber orientation and distribution: the first one is focused on the experiments with SFRSCC specimens and further x-ray computed tomography technique (Sect. 2.1), the second approach are "simulation experiments" and are concerned with a transparent polymer mixture as replacement for the opaque concrete (Sect. 2.2).

2.1 Concrete Experiments: Fresh Concrete

In this study, we examined a slab of steel fiber reinforced self-compacting concrete (SFRSCC) of dimensions 90 cm × 60 cm × 20 cm. The casting process was conducted from a hopper positioned in the middle of the edge of the formwork. The point of slump was located at one of the 60 cm edges, see Fig. 1. The formwork was sprayed with a thin layer of oil before casting to simplify the further demolding process. One side of the form was equipped with lifting anchors to simplify the lifting of the slab.

The mix proportion of self-compacting concrete have been chosen by the producer. Hooked end steel fibers were gradually added to the self-compacting concrete and mixed. The fiber volume ratio was 0.5%. The fiber length and the fiber diameter were

Fig. 1 Casting of a small plate using a bucket and half-funnel

Table 1 Fiber concrete data

Casting day		30.09.16
Slump (diameter) of concrete		75 cm
Temperature		20 °C
Density and Compressive strength at 3 day	2470 kg/m^3	48.2 MPa
Density and Compressive strength at 7 day	2460 kg/m^3	61.1 MPa
Density and Compressive strength at 28 day	2470 kg/m^3 2470 kg/m^3	71.2 MPa 67.5 MPa

60 mm and 0.75 mm, respectively; the fiber properties are summarized in Table 2. The properties of the hardened concrete are given in Table 1.

After hardening in order to make an orientation analysis with x-ray CT possible, the slab was cut into three beams of 90 cm × 19.5 cm × 19.5 cm using a diamond saw, as indicated in Fig. 2. The arrows represent the flow direction of concrete. One of the beams contains the lifting anchors.

Table 2 Data of used fibers: Semtu WireFib 80/60; amount used 25 kg/m^3

Length	60 mm
Diameter	0.75 mm
Aspect ratio	80
Number of fibers/kg n	4600
Tensile strength	>1000 MPa
Coating	uncoated
Steel quality	EN 10016-2 C9

2.2 *Simulation Experiments: Transparent Replacement Matrix*

The method used in this research is similar to that in [8, 29]. It is based on replacing of SFRSCC by transparent polymer mixture, allowing observation of fiber orientation and distribution during and immediately after the casting process. For this experiment, a mixture was produced according to the manufacturer's guidelines of the Sodium PVM/MA Stabilizer. The experimental mixture consists of a stabilizer powder, water and sodium hydroxide (NaOH). The mixture is produced in two main stages. In the first stage, the stabilizer powder is dissolved in the water at room temperature and the obtained mixture is neutralized with NaOH. The mixture should be regularly shaken/stirred to achieve a better homogeneity and dispersion with minimal amount of air voids and bubbles. The second stage ensures higher transparency and gel viscosity of the experimental fluid. Finally, after multiple adding of NaOH and frequent shaking, the PH of the obtained mixture was equal to 7. Then the mixture was left for a couple of days to achieve better viscosity and transparency.

Before the casting experiments, the rheological properties of the mixture were tested by means of the standardized slump flow test. The suitable mixture should represent a homogeneous and visco-plastic fluid (shows Bingham plastic behavior).

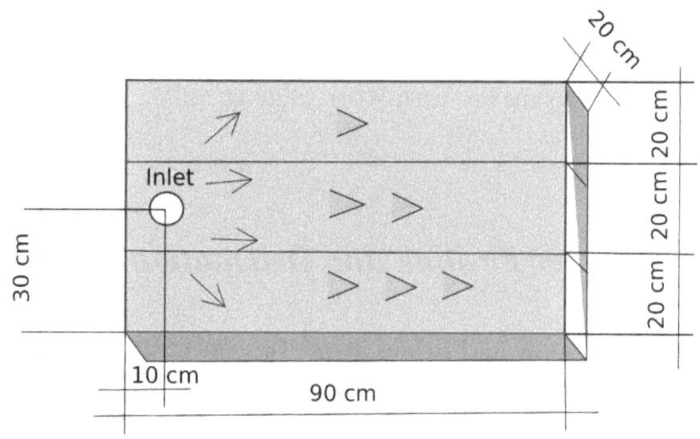

Fig. 2 Layout of beam-cutting of small plate

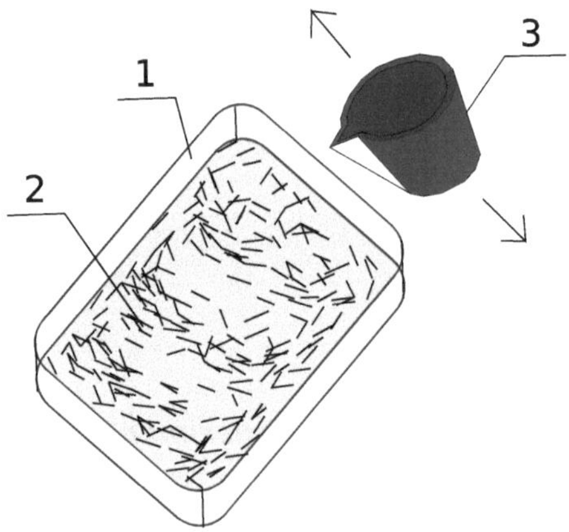

Fig. 3 Casting using a bucket. 1: plastic box, 2: polymer with fibers, 3: 12 l bucket

When the mixture was ready, the matchsticks fibers were gradually added in a proportion of 3%. The length of the matchstick fibers was 50 mm and the aspect ratio (fiber length/fiber diameter) was 25.

The main aim of the casting experiments was to investigate the correlation between the casting conditions—e.g. formwork surface quality—fiber orientation and distribution experimentally. For that purpose, the following series of experiments with different casting conditions and formwork surface quality was performed:

1. Experiment with rough formwork surface and using a fixed pipe located in the middle of the formwork edge;
2. Experiment with smooth formwork surface and using a fixed pipe located in the middle of the formwork edge;
3. Experiment with rough formwork surface and using a bucket located in the middle of the formwork edge;
4. Experiment with smooth formwork surface and using a bucket located in the middle of the formwork edge;
5. Experiment with smooth formwork surface and using a bucket moving back and forth along the formwork edge;
6. Experiment with rough formwork surface and using a bucket moving back and forth along the formwork edge (Fig. 3).

2.3 *Fiber Position and Orientation Measurement*

Fiber position and orientations in real world are 3D, but here we only consider a thin layer, the bottom layer and therefore use a 2D description, namely x and y coordinates, and one angle θ, the angle between the fiber and the x-axis, counted counter-clockwise.

To describe the fiber orientation[1] and alignment[2] mathematically, several measures are common, the orientation factor, orientation number which are closely related and based on a pre-set cross-section or preference direction. Mathematically more sophisticated measures exist in the form of the orientation or alignment tensor and the scalar order parameter [30–32].

The orientation factor relates the average fiber orientation to the number of fibers crossing a given cross-section, e.g. a crack-plane:

$$\alpha = \frac{N_f A_f}{A_c v_f} , \tag{1}$$

where A_c is the cross-section area of the concrete under consideration, N_f is the number of fibers in the cross-section A_c, A_f is the fiber cross-section and v_f is the volume fraction of the fibers.

The orientation number calculates the average projected length of a fiber onto a given direction, e.g. the normal to the crack-plane:

$$\eta = \frac{1}{N} \sum_{i=1}^{N} \cos\theta_i , \tag{2}$$

where θ_i is the angle between the i-th fiber and a given axis, e.g. the x-axis.

The tensor-based orientation and alignment measures are based on the use of spherical harmonical functions and an orientation *distribution* on a unit sphere, for details, see [30–32]. The scalar order parameter is given as:

$$S = \frac{< d\cos^2(\phi_i) - 1 >_i}{d - 1} , \tag{3}$$

$$\text{in 3D:} \quad S = \frac{< 3\cos^2(\phi_i) - 1 >_{i=1 \ldots N}}{2} , \tag{4}$$

$$\text{in 2D:} \quad S = \frac{< 2\cos^2(\phi_i) - 1 >_{i=1 \ldots N}}{1} , \tag{5}$$

where $< \cdot >$ denotes the ensemble average over all N fibers, d is the space dimension and ϕ_i is the angle between the i-th fiber and the average orientation. The average orientation can be obtained from the eigenvectors and eigenvalues of the second order orientation tensor.

The scalar order parameter is independent of a pre-defined direction, as it describes how well the fibers are aligned with each other. The eigenvectors of the alignment tensor describe in which direction the fibers are oriented.

Fiber orientations and their alignment can be visualized by orientation ellipses, whose major axes are oriented according to the eigenvectors and the scaling of the

[1]Orientation as in oriented in a (certain) direction.

[2]Alignment as in aligned with each other.

axes is according to the eigenvalues of the second order orientation or alignment tensor.

$$\frac{(e_1^T x)^2}{\left(\frac{1}{\sqrt{\lambda_1}}\right)^2} + \frac{(e_2^T x)^2}{\left(\frac{1}{\sqrt{\lambda_2}}\right)^2} = 1 \tag{6}$$

or, with all ellipses having the same major axis (assuming λ_1 is the smaller eigenvalue)

$$\frac{(e_1^T x)^2}{1} + \frac{(e_2^T x)^2}{\left(\sqrt{\frac{\lambda_2}{\lambda_1}}\right)^2} = 1\,. \tag{7}$$

3 Results of Casting Experiments

In this section the results of the analysis of the spatial and orientational distribution of the fibers for the SFRC and for the "simulation" experiments with the polymer matrix will be presented. The fiber orientations have been obtained from images of the cast material. In the case of the SFRC, this has been x-ray CT images, and in the case of the polymer matrix, photos have been taken after the casting. In both cases, the fibers have been analyzed using the measurement tool from ImageJ [33].

3.1 *Casting of a Small SFRC Plate: Fiber Orientations and Spatial Distribution*

The three beams have been scanned using a medical x-ray CT scanner. The bottom layer of the volume images was then thresholded to show the fibers. In the resulting image, the fibers were measured using the measurement tool of ImageJ with manual identification of the fibers. This was necessary, as automatic fiber analysis was difficult due to the low resolution of the scanner compared to the fiber diameter and the large sample diameter, which had caused noise and artifacts together with a strong cupping effect.

The orientation and position analysis of the received data was performed using self-developed scripts [34] in R [35] calculating the orientation tensor and its eigenvalues and eigenvectors and finally plotting [36] the orientation ellipses.

Figure 4 shows the results of the orientation and spatial distribution analysis. As can be seen, the orientation distribution resembles one predicted for a slippery formwork [15]. Especially the fibers are oriented mostly perpendicular to the flow velocity.

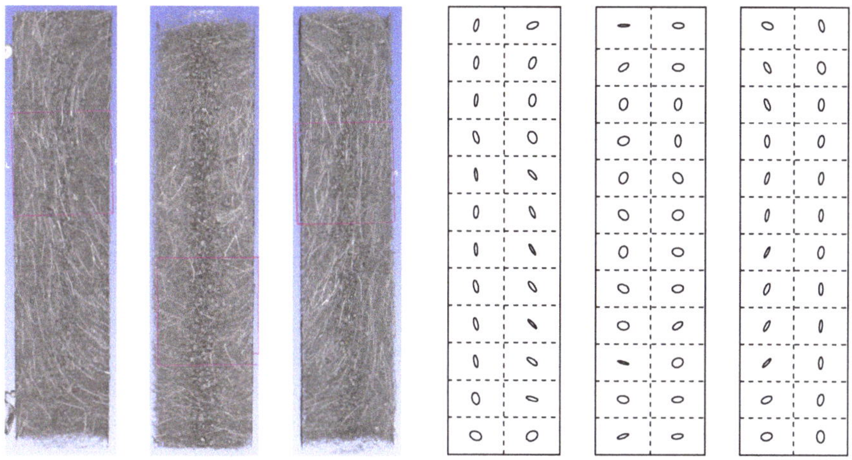

(a) Beam 3 (b) beam 2 (c) Beam 1 (d) Beam 3 (e) Beam 2 (f) Beam 1

21	9
15	14
7	9
19	14
15	7
15	13
16	11
14	13
12	14
11	15
9	14
4	5

(g) Beam 3

7	12
13	13
21	19
14	18
16	17
16	20
16	14
13	16
13	14
9	18
11	12
9	7

(h) Beam 2

18	15
18	11
6	9
13	11
11	9
18	8
14	10
17	15
14	12
13	15
13	12
7	8

(i) Beam 1

Fig. 4 Bottom layer of the beams in x-ray CT. Casting point was at the top of the middle beam. The fibers are mostly oriented parallel to the flow front, meaning perpendicular to the flow velocity

3.2 *Replacement Castings: Fiber Orientations and Spatial Distribution*

Preliminary observations were obtained immediately after each experiment, which show that the fiber orientation is mainly perpendicular to the flow direction but their spatial distribution is quite different.

After each series of experiments, we obtained the images from above, which were subjected to further image analysis. A typical image obtained during the experiment is presented in Table 3.

The results of the spatial and orientational analysis for the different casting schemes are presented in Figs. 5, 6, 7 and 8. As can be seen, the fibers are mainly oriented perpendicular to the flow direction. The dependence of the orientational

Table 3 Matrix of experiments

bottom	slow	fast
smooth dry	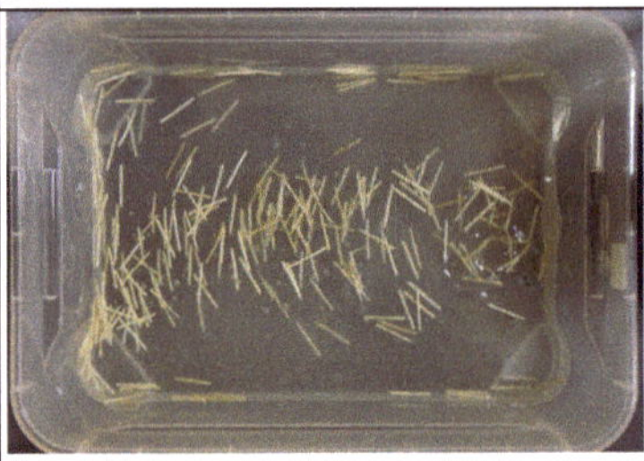	
smooth wet	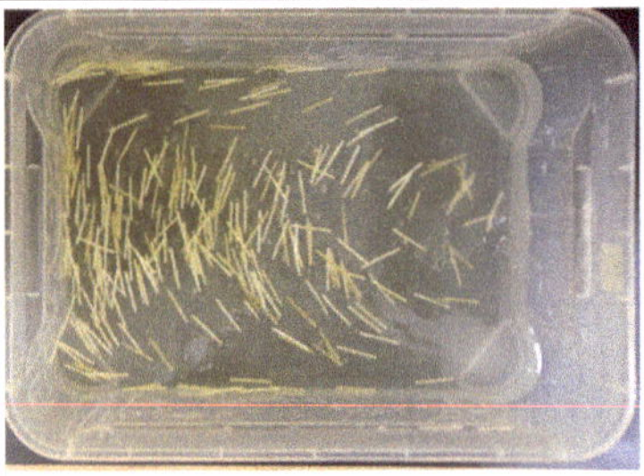	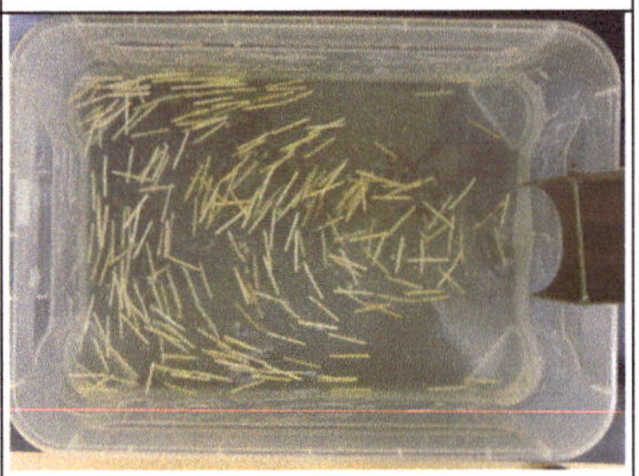
rough dry		

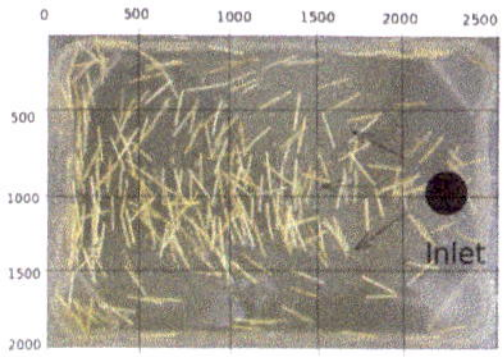

(a) Photo of fibers with cell division

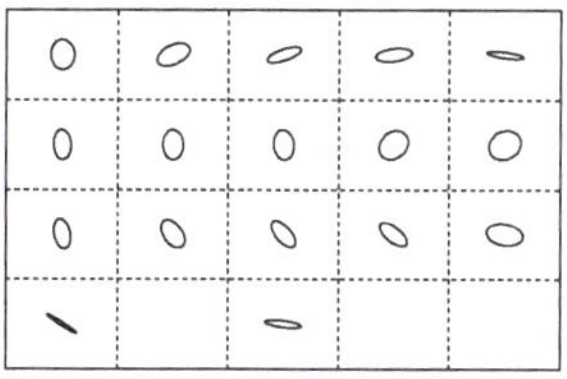

(b) Fiber test: orientation ellipses per cell

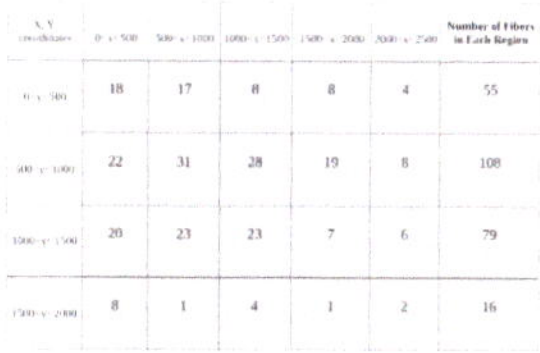

X, Y coordinates	0 < x < 500	500 < x < 1000	1000 < x < 1500	1500 < x < 2000	2000 < x < 2500	Number of Fibers in Each Region
0 < y < 500	18	17	8	8	4	55
500 < y < 1000	22	31	28	19	8	108
1000 < y < 1500	20	23	23	7	6	79
1500 < y < 2000	8	1	4	1	2	16

(c) Fiber count per area

Fig. 5 Fiber orientations, smooth bottom, fast casting

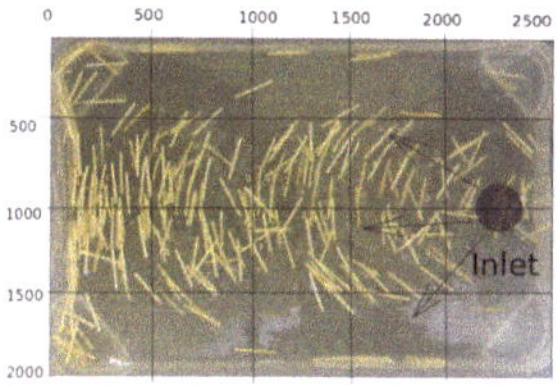

(a) Photo of fibers with cell division

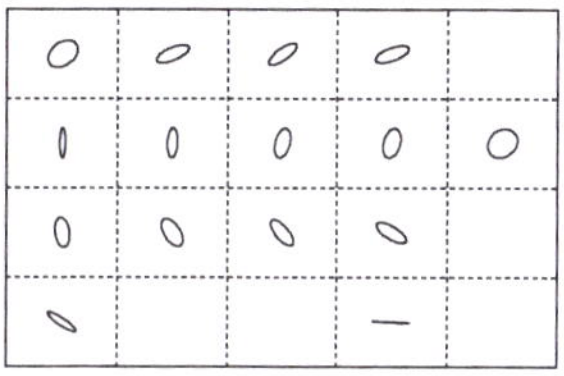

(b) Fiber test: orientation ellipses per cell

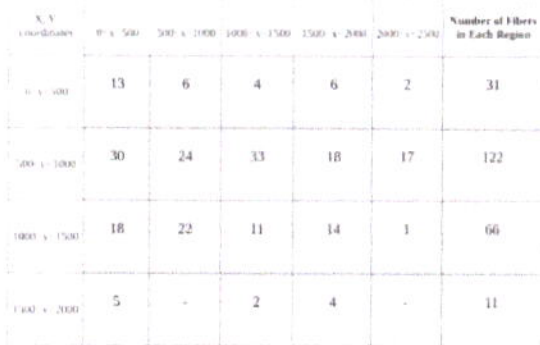

X, Y coordinates	0 < x < 500	500 < x < 1000	1000 < x < 1500	1500 < x < 2000	2000 < x < 2500	Number of Fibers in Each Region
0 < y < 500	13	6	4	6	2	31
500 < y < 1000	30	24	33	18	17	122
1000 < y < 1500	18	22	11	14	1	66
1500 < y < 2000	5	-	2	4	-	11

(c) Fiber count per area

Fig. 6 Fiber orientations, smooth bottom, slow casting

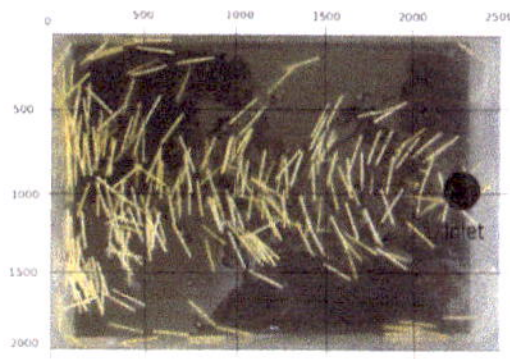

(a) Photo of fibers with cell division

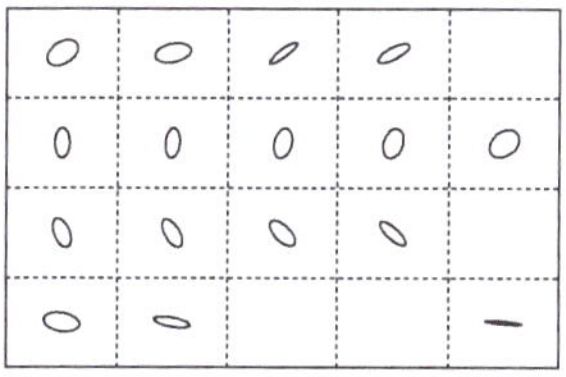

(b) Fiber test: orientation ellipses per cell

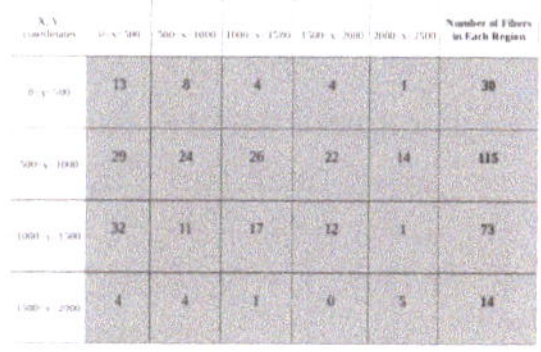

X, Y coordinates	0 < x < 500	500 < x < 1000	1000 < x < 1500	1500 < x < 2000	2000 < x < 2500	Number of Fibers in Each Region
0 < y < 500	13	8	4	4	1	30
500 < y < 1000	29	24	26	22	14	115
1000 < y < 1500	32	11	17	12	1	73
1500 < y < 2000	4	4	1	0	5	14

(c) Fiber count per area

Fig. 7 Fiber orientations, rough, slow

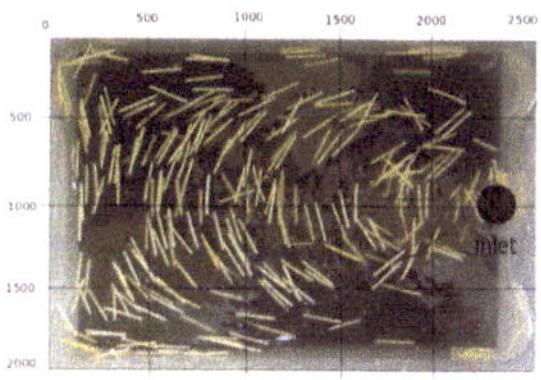

(a) Photo of fibers with cell division

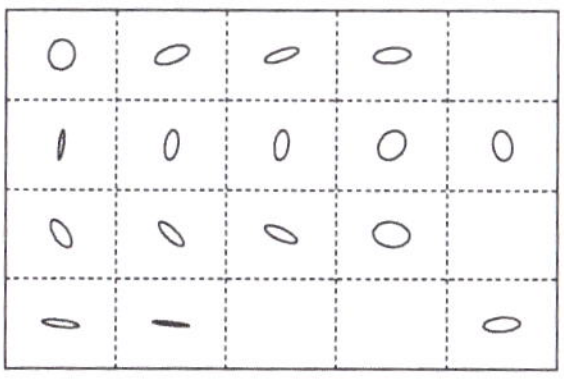

(b) Fiber test: orientation ellipses per cell

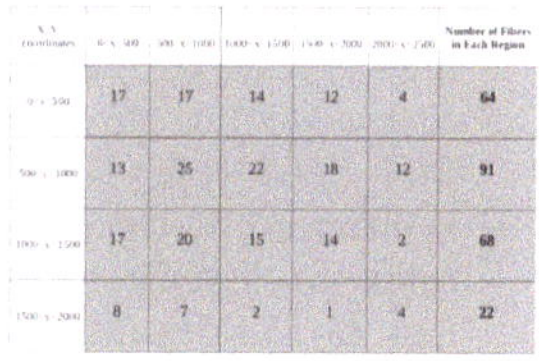

X, Y coordinates	0 < x < 500	500 < x < 1000	1000 < x < 1500	1500 < x < 2000	2000 < x < 2500	Number of Fibers in Each Region
0 < y < 500	17	17	14	12	4	64
500 < y < 1000	13	25	22	18	12	91
1000 < y < 1500	17	20	15	14	2	68
1500 < y < 2000	8	7	2	1	4	22

(c) Fiber count per area

Fig. 8 Fiber orientations, rough bottom, fast casting

distribution on the smoothness of the formwork bottom is inconclusive. The spatial distribution show a dependence on the casting velocity, with the fibers concentrating in a narrower channel in cases of very slow casting.

4 Conclusion

Several factors influencing the fiber orientations and placement in fiber reinforced materials have been experimentally investigated. The experiments showed a surprising influence of the casting velocity on the spatial distribution of fibers, while only small influence on the orientational distribution. The roughness of the formwork bottom showed some influence on the fiber orientations but no significant influence on the spatial dispersion. While some influences have been identified in this initial study, further investigations are necessary.

Acknowledgements The authors gratefully acknowledge the funding by the Estonian Research Council by the exploratory research grant PUT1146.

We also thank Maria Kremsreiter who helped during her Erasmus+ internship at the Institute of Cybernetics. Therefore: With the support of the Erasmus+ programme of the European Union. Thanks to E-Betoonelement, especially Aare Lessuk, Rasmus-R. Marjapuu and Sergei Graf, for preparing the experiment plate.

References

1. Beissel, R.A., Lim, H.: Self Compacting Concrete: Modern Concrete and Admixture Technology, pp. 607–612 (2000)
2. Brouwers, H.J.H., Radix, H.J.: Self-compacting concrete: theoretical and experimental study. Cem. Concr. Res. **35**(11), 2116–2136 (2005)
3. Nawy, E.G., Edward, G.: Fundamentals of High-performance Concrete. Wiley (2001)
4. Balaguru, P.N., Shah, S.P.: Fiber-reinforced cement composites (1992)
5. Lawler, J.S., Zampini, D., Shah, S.P.: Permeability of cracked hybrid fiber-reinforced mortar under load. Mater. J. **99**(4), 379–385 (2002)
6. Voigt, T., Van Bui, K., Shah, S.P.: Drying shrinkage of concrete reinforced with fibers and welded-wire fabric. Mater. J. **101**(2), 233–241 (2004)
7. Mesbah, H.A., Buyle-Bodin, F.: Efficiency of polypropylene and metallic fibres on control of shrinkage and cracking of recycled aggregate mortars. Constr. Build. Mater. **13**(8), 439–447 (1999)
8. Boulekbache, B., Hamrat, M., Chemrouk, M., Amziane, S.: Flowability of fibre-reinforced concrete and its effect on the mechanical properties of the material. Constr. Build. Mater. **24**(9), 1664–1671 (2010)
9. Ponikiewski, T., Gołaszewski, J., Rudzki, M., Bugdol, M.: Determination of steel fibres distribution in self-compacting concrete beams using x-ray computed tomography. Arch. Civ. Mech. Eng. **15**, 558–568 (2015)
10. Liu, J., Sun, W., Miao, C., Liu, J., Li, C.: Assessment of fiber distribution in steel fiber mortar using image analysis. J. Wuhan Univ. Technol. Mater. Sci. Ed. **27**, 166–171 (2012)
11. Stähli, P., Custer, R., van Mier, J.G.M.: On flow properties, fibre distribution, fibre orientation and flexural behaviour of FRC. Mater. Struct. **41**(1), 189–196 (2008)
12. Švec, O., Žirgulis, G., Bolander, J.E., Stang, H.: Influence of formwork surface on the orientation of steel fibres within self-compacting concrete and on the mechanical properties of cast structural elements. Cem. Concr. Compos. **50**, 60–72 (2014)
13. Soroushian, P., Lee, C.-D.: Distribution and orientation of fibers in steel fiber reinforced concrete. Mater. J. **87**, 433–439 (1990)

14. Grigaliunas, P., Kringelis, T.: SCC flow induced steel fiber distribution and orientation. Non-destructive inductive method. In: 11th European Conference on Non-Destructive Testing, Prague, Czech Republic (2014)
15. Herrmann, H., Lees, A.: On the influence of the rheological boundary conditions on the fibre orientations in the production of steel fibre reinforced concrete elements. Proc. Est. Acad. Sci. **65**(4), 408–413 (2016). Open-Access CC-BY-NC 4.0
16. Promentilla, M.A.B., Sugiyama, T., Shimura, K.: Threedimensional imaging of cement-based materials with x-ray tomographic microscopy: visualization and quantification. In: International Conference on Microstructure Related Durability of Cementitious Composites, vol. 61, pp. 1357–1366 (2008)
17. Liu, J., Li, C., Liu, J., Cui, G., Yang, Z.: Study on 3D spatial distribution of steel fibers in fiber reinforced cementitious composites through micro-CT technique. Constr. Build. Mater. **48**, 656–661 (2013)
18. Pastorelli, E., Herrmann, H.: Time-efficient automated analysis for fibre orientations in steel fibre reinforced concrete. Proc. Est. Acad. Sci. **65**(1), 28–36 (2016)
19. Herrmann, H., Pastorelli, E., Kallonen, A., Suuronen, J.-P.: Methods for fibre orientation analysis of x-ray tomography images of steel fibre reinforced concrete (SFRC). J. Mater. Sci. **51**(8), 3772–3783 (2016)
20. Suuronen, J.-P., Kallonen, A., Eik, M., Puttonen, J., Serimaa, R., Herrmann, H.: Analysis of short fibres orientation in steel fibre reinforced concrete (SFRC) using x-ray tomography. J. Mater. Sci. **48**(3), 1358–1367 (2013)
21. Ferrara, L., Faifer, M., Toscani, S.: A magnetic method for non destructive monitoring of fiber dispersion and orientation in steel fiber reinforced cementitious composites–Part 1: method calibration. Mater. Struct., 1–15 (2011)
22. Karhunen, K., Seppänen, A., Lehikoinen, A., Monteiro, P.J.M., Kaipio, J.P.: Electrical resistance tomography imaging of concrete. Cem. Concr. Res. **40**, 137–145 (2010)
23. Torrents, J.M., Blanco, A., Pujadas, P., Aguado, A., Juan-Garcia, P., Sánchez-Moragues, M.Á.: Inductive method for assessing the amount and orientation of steel fibers in concrete. Materi. Struct. **45**, 1577–1592 (2012)
24. Schickert, M.: Progress in ultrasonic imaging of concrete. Mater. Struct. **38**, 807–815 (2005)
25. Aggelis, D.G., Soulioti, D., Barkoula, N.M., Paipetis, A.S., Matikas, T.E., Shiotani, T.: Acoustic emission monitoring of steel-fiber reinforced concrete beams under bending. Prog AE **14**, 287–294 (2008)
26. Keru, W., Chen, B., Yao, W.: Study on the ae characteristics of fracture process of mortar, concrete and steel-fiber-reinforced concrete beams. Cem. Concr. Res. **30**(9), 1495–1500 (2000)
27. Soulioti, D., Barkoula, N.M., Paipetis, A., Matikas, T.E., Shiotani, T., Aggelis, D.G.: Acoustic emission behavior of steel fibre reinforced concrete under bending. Constr. Build. Mater. **23**(12), 3532–3536 (2009)
28. Roussel, N., Gram, A., Cremonesi, M., Ferrara, L., Krenzer, K., Mechtcherine, V., Shyshko, S., Skocec, J., Spangenberg, J., Svec, O., Thrane, L.N., Vasilic, K.: Numerical simulations of concrete flow: a benchmark comparison. Cem. Concr. Res. **79**, 265–271 (2016)
29. Svec, O., Skocek, J., Olesen, J.F., Stang, H.: Fibre reinforced self-compacting concrete flow simulations in comparison with l-box experiments using carbopol. In: 8th Rilem International Symposium on Fibre Reinforced Concrete (2012)
30. Ehrentraut, H., Muschik, W.: On symmetric irreducible tensors in d-dimensions. ARI Int. J. Phys. Eng. Sci. **51**(2), 149–159 (1998)
31. Herrmann, H., Beddig, M.: Tensor series expansion of a spherical function for use in constitutive theory of materials containing orientable particles. Proc. Est. Acad. Sci. **67**(1), 73–92 (2018). Open-Access CC-BY-NC 4.0
32. Eik, M., Puttonen, J., Herrmann, H.: The effect of approximation accuracy of the orientation distribution function on the elastic properties of short fibre reinforced composites. Compos. Struct. **148**, 12–18 (2016)
33. Abràmoff, M.D., Magalhães, P.J., Ram, S.J.: Image processing with ImageJ. Biophotonics Int. **11**(7), 36–42 (2004)

34. Herrmann, H.: Alignment tensor package for R (2016). http://bitbucket.org/vispar/alignment
35. R Development Core Team: R: A Language and Environment for Statistical Computing. R Foundation for Statistical Computing, Vienna, Austria (2011). ISBN 3-900051-07-0
36. Lemon, J.: Plotrix: a package in the red light district of R. R-News **6**(4), 8–12 (2006)

Image Data Processing to Obtain Fibre Orientation in Fibre-Reinforced Elements Using Computed Tomography Scan

Jesús Mínguez, Miguel A. Vicente and Dorys C. González

Abstract Computed tomography (CT) technique is of increasing interest in research related to concrete technology. This technology provides the possibility of visualize the internal structure of concrete, including pores, cracks, aggregates and fibres. In this paper, the CT scan is used to determine the position and orientation of the fibres in case of steel fibre reinforced high strength concrete elements (SFRHSC). This paper shows a home-made numerical procedure, automated through a MATLAB routine, which enables, fast and reliable, get the orientation of each and every one of the fibres and their center of gravity. The procedure shown can be used with any type of fibre reinforced material, with the only restriction that a wide difference between density of fibres and density of matrix is needed. The algorithm is simple and robust. The result is a fast algorithm and a routine easy to use. In addition, the validation tests show that the error is almost zero.

1 Introduction

During the last years there have been a significant increase in the use of fibres (metallic and/or nonmetallic) as a substitute, in whole or in part, for rebar in high performance concrete elements. This solution provides some interesting advantages. The most important is the manufacturing cost reduction. In addition, fibres provides, in general, a significant increase of the ductility of hardened concrete. This property

J. Mínguez (✉) · M. A. Vicente · D. C. González
Department of Civil Engineering, University of Burgos, Burgos, Spain
e-mail: jminguez@ubu.es

M. A. Vicente
e-mail: mvicente@ubu.es

D. C. González
e-mail: dgonzalez@ubu.es

H. Herrmann and J. Schnell (eds.), *Short Fibre Reinforced Cementitious Composites and Ceramics*, Advanced Structured Materials 95,
https://doi.org/10.1007/978-3-030-00868-0_8

is highly demanded, especially in the case of concrete elements that are subjected to cyclic loading.

The most relevant disadvantage is that fibres reduce the workability of concrete. The reduction strongly depends on the type of fibre and quantity, but, in general, a relevant reduction of the workability can be observed. In addition, an increase of the scatter of the workability can be observed.

There are many factors affecting the efficiency of the fibres. The most relevant are the bond with the cement paste, their slenderness, the degree of dispersion in the concrete mass and, in particular, their orientation.

All the mechanical parameters of hardened concrete are influenced by fibres: compression strength, Young's modulus, Poisson ratio, fracture energy, etc. However, the most affected factor is the tensile strength.

Regarding to this last parameter, it can be highlighted that it is strongly influenced by fibre shape, amount and orientation (e.g., [6–9, 11, 13, 20, 27]).

From a structural point of view, the fibres must be aligned with the principal tensile stress directions. If it happens, the efficiency of the fibres reaches the maximum and, in consequence, the minimum amount of fibres is required.

Up to date, the orientation of fibres inside concrete mass is a stochastic process, which cannot be controlled. The orientation of the fibres within a concrete mass depends on many factors, including the shape of the formwork, the concrete pouring process and the relative size between fibres and aggregates. Many research efforts have been carried out during the last years in order to understand and manage the orientation of fibres inside the concrete [10, 17, 18, 22, 28].

In recent years, several methods have been developed for determining the dominant orientation of the fibres. These techniques can be classified into two main groups: indirect methods and methods using CT-Scan technology. These one can also be classified into semi-direct methods and direct methods. Next, the most relevant of these methods are explained.

1.1 Indirect Methods

Indirect methods measure other material parameters that are related to the fibre orientation. Some of the most typical approaches are discussed below.

The AC Impedance Spectroscopy method [14, 15, 23, 25, 26] measures the impedance of an induced electrical current throughout the specimens, by using two electrodes placed on the test specimen. Impedance values vary with the path of electric current inside the specimen and it is strongly related to the overall amount of fibres and orientation.

Similar to the previous one is the Impedance Over-Frequency technique [3]. Compared to conventional AC Impedance Spectrometry, less expensive equipment is required. Additionally, it can not only be adopted for laboratory tests but also for the monitoring of elements already in place.

Other technique is the Microwave Reflectometry. This method allows measurement of the fibre concentration [1]. This principle of working is that steel fibres inside a homogeneous concrete matrix affects the effective permittivity of the mixture. This property depends on the volume fraction of the fibres and their aspect ratios.

The method of Magnetic Fields [4] is the most commonly used indirect method. It is based on the fact that concrete shows a notably higher magnetic permeability than fibres, under a magnetic field. The effective permeability parameter is used to determine the concentration of fibres and their orientation.

All these methods show two main limitations. First, the results are notably sensitive to the testing procedure; in consequence, different technicians may obtain different results. Second, these methods can only be used with ferromagnetic fibres.

1.2 Methods Using CT-Scan Technology

The indirect methods do not provide accurate information about the fibre orientation, only general information about dominant direction. Furthermore, they are not able to provide singularities inside the specimen.

Up to date, computed tomography (CT) technology is the only one which allows visualization of the exact position of each fibre within a hardened concrete mass.

1.2.1 CT Technology

Computed Tomography (CT) is a non-destructive technique used to analyze the internal microstructure of materials, based on the property of x-rays. The tomography equipment is composed of an emitter, which emits a ray at a given intensity, and a detector, which registers the reception intensity of the ray. In the analysis, the object revolves in front of the apparatus, consisting of the emitter, emitting rays in all directions on the plane, and the detector. Post-processing of the signal to produce attenuation-corrected images, which coincide with the measurement of attenuation, means that the density of each point of the specimen under study may be determined. This process is repeated for different sections of the specimen, thereby obtaining tridimensional (tomographic) information. Alternatively, a conic beam of X-rays can be emitted that are collected on a flat detector. In this case, only the specimen has to revolve and relative displacement between the emitter–detector apparatus and the specimen is unnecessary.

Nowadays, the use of this technology is very wide, from the most conventional medicine to soil mechanics or concrete, passing through paleontology, heritage, asphalts, metals, composites, etc. [24].

Both semi-direct and direct methods use the data obtained from the CT-Scan, but in a different way, i.e., with a different post-processing protocol. Next, both methods are described.

1.2.2 Semi-direct Methods

The principle of working of the semi-direct methods is to analyse fibre orientations in an indirect way through analyzing the positions of the pixels belonging to fibres inside the specimen.

It is worth highlighting the work carried out by Schnell et al. [19]. They have developed a procedure for determining the orientation of fibres through the determination of their apparent length in the three space directions through the two-dimensional analysis of the information in each of the three space directions.

In a similar way, Stroeven and Hu [21] propose the use of the stereology. This technique is based on fundamental principles of geometry and statistics. The main hypothesis of stereology is that some 3-D quantities can be determined without 3-D reconstruction.

1.2.3 Direct Methods

The semi-direct methods show a higher accuracy than the indirect ones, but they still show some deficiencies. They are not able to define the exact orientations of the fibres individually. In consequence, it is not possible to determine multiple dominant directions.

Direct methods also use the information obtained from computed tomography scans. Once the 3D image of the especimen is obtained, and the voxels belonging to fibre are identified and extracted, it is possible to cluster them to obtain each individual fibre. Next, it is possible to determine the spatial orientation of each fibre.

This method provides the enormous advantage that the orientation of each individual fibre can be obtained. Next, the dominant orientation can be defined using statistical tools. Using this technology it is possible to detect whether there are two or more dominant orientations inside the specimen.

Moreover, it is possible to extract smaller volumes from the whole 3D model of the specimen, and analyse the orientation of the fibres in these smaller volumes. Thus, for example, it is possible to identify differences in fibre orientations in the different regions of the specimen (e.g. between the border region and the inner one).

The main limitation in the use of this technology is the development of post-processing algorithms, able to manage the huge information provided by the CT-Scan.

Some interesting works have been developed in the last years showing post-processing algorithms.

An interesting work in this group of methods was carried out by Krause et al. [12]. This one proposes a mathematical technique to obtain the fibre orientations in composites (including fibre reinforced concrete). This technique is based on the structure tensor, which evaluates the average gradient direction in the nearby of a point x. In addition, the structure tensor can be used to define a direction in which the image information is constant, if such a direction exists. This leads to the computation of fibre orientation.

Table 1 Mix proportions

Cement (kg/m^3)	Water (kg/m^3)	Water reducer (kg/m^3)	Silica fume (kg/m^3)	Fine aggregate (kg/m^3)	Coarse aggregate (kg/m^3)	Fibre
400	125	14	6	800	1,080	1% by Volume

Pastorelli, Herrmann and coworkers [5, 16], have developed an interesting protocol based on the Hessian matrix. This procedure improves some numerical problems shown by previous ones; for example, it solves properly the common problem of separating touching fibres. On the contrary, the solution is requires high computer capacity.

This paper shows a new procedure to obtain the exact orientation of each fibre inside the concrete specimen. This method shows two advantages above other methods (mainly the indirect methods). The first one is that the test is more accurate. The second one is that the test is much easier and faster because it does not require large laboratory facilities but only one CT-Scan.

The procedure shown can be applied to any type of fibre reinforced composites, whenever there is a wide difference between density of fibres and matrix. Otherwise CT-scan is not able to identify separately the pixels belonging to fibres from the pixels belonging to matrix.

2 Experimental Investigation

A set of eight steel-fibre reinforced concrete cylinders 100 mm diameter, 200 mm height have been built to validate the procedure to obtain the fibre orientation. Table 1 shows the concrete proportions.

In this case, steel hooked-end fibres, 35 mm length, 1 mm diameter were used. The quantity of fibres used was 1% by volume.

Each specimen was scanned using a CT-Scan model Y.CT-Compact unit from YXLON (Hamburg, Germany), property of the University of Burgos (Spain) (Fig. 1). It is equipped with a 450-kV X-ray source and a line detector. This facility records sectional radiographs of the specimen at different heights (Fig. 2a), spaced 0.1 mm or more.

The characteristics of each image are shown in Table 2.

The result of this analysis is a set of 2-D 8-bits grey-scaled slices, where each grey tone corresponds to a density value of the pixel such that the denser regions are shown in lighter greys, while less dense regions are shown in darker grey. A total of 256 grey tones are considered. Pixels belonging to steel fibres show lighter shades of grey.

Fig. 1 Testing specimen inside the CT scan

The distance between slices is 0.5 mm, which is less than the thickness of the fibres so that no fibres are "hidden" between two planes.

The numerical procedure have been implemented using a MATLAB (Mathworks, USA) routine, specially developed by the authors. This MATLAB routine is composed by several sub-routines.

A first MATLAB sub-routine identifies the pixels belonging to fibres in each slice by considering only the grey tone pixels that correspond to the fibres and discarding the others. Once the pixels belonging to fibres are identified, it is possible to build a 3D image including the fibres only. The procedure to identify the pixels belonging to fibre is through the definition of a grey-threshold value in such a way that lighter grey pixels belong to fibres and darker grey pixels belong to matrix (Fig. 3).

Once the pixels corresponding to the fibres are extracted, they are exported as nodes to a data file. Each pixel has the following information: X, Y and Z coordinates and grey-scale color (Fig. 2b)

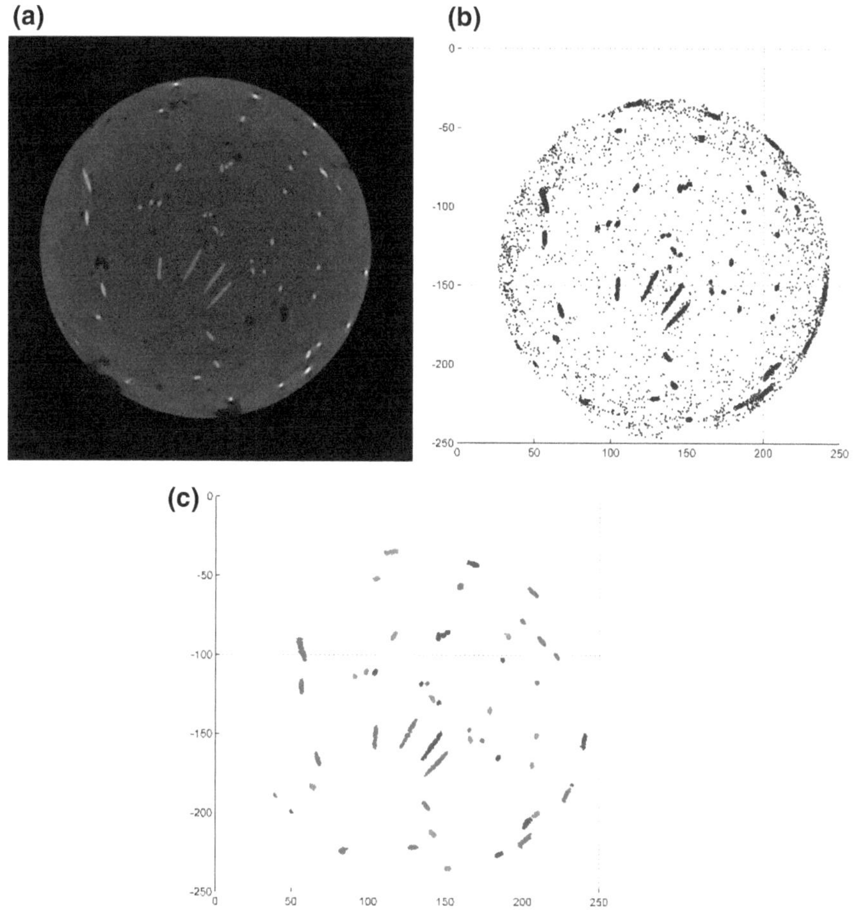

Fig. 2 Segmentation and core grouping procedure. **a** Example of a CT scan cross section image. **b** Fibre separated from concrete matrix. **c** Image filtered and fibre grouping

Table 2 Scanned image characteristics

CT: Number of projections	720
DR: Number of lines	440
Multi-Scan: Number of slices	408
Multi Slice: Distance between two slices [mm]	0.5
Magnification (Object centre)	1.33
Pixel size [mm]	0.124

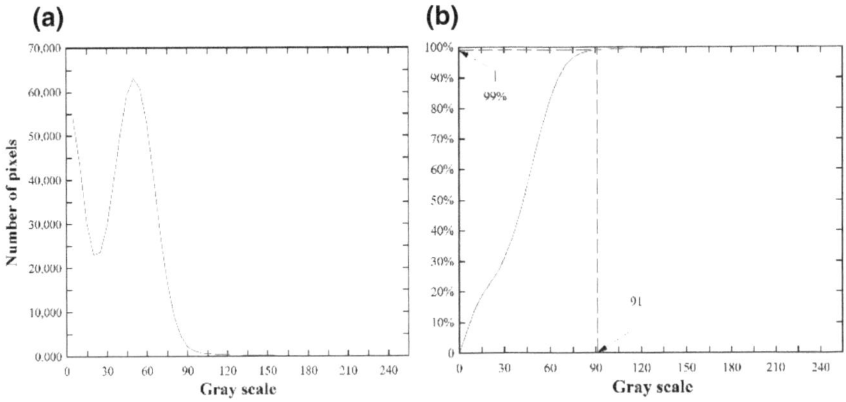

Fig. 3 Grayscale ordinary histogram (**a**) and cumulative histogram (**b**)

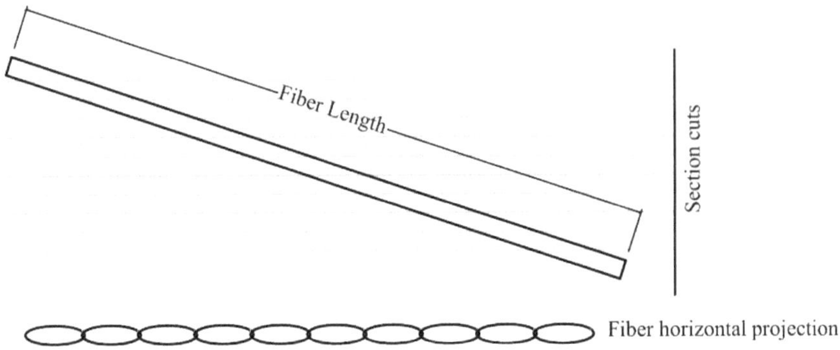

Fig. 4 Fibre section cut horizontal projection

This data file is processed by a second MATLAB sub-routine, which identifies and groups those nodes belonging to the same fibre. The procedure is divided into several phases. First, pixels belonging to the same fibre are grouped, inside each slice. The clustering algorithm uses a limit tolerance related to the pixel size in such a way that two pixels that they are closer than the tolerance belongs to the same fibre (Fig. 2c).

Next, the total amount of pixels belonging to the same group is checked and if this number is small than the expected they are discarded. This is a filtering process in order to avoid numerical noise.

Then the groups of pixels belonging to different slices are clustered. To do it, a search algorithm with the same tolerance as the previous phase is developed. In this case, if the horizontal projection distance between two points belonging to different groups in consecutive cross sections is less than the tolerance limit, then they belong to the same fibre and they are grouped into a fibre (Fig. 4). This process is repeated in the following cross sections until there are no pixels to group into a fibre.

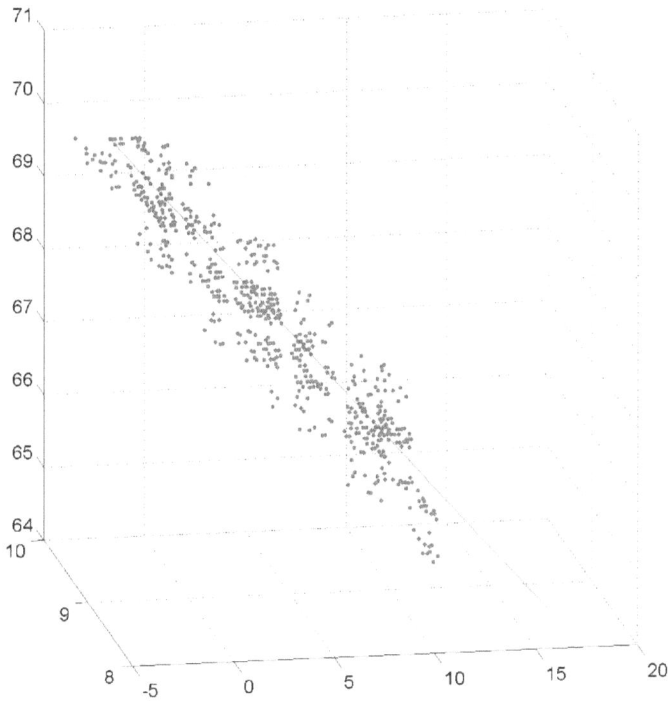

Fig. 5 Linear regression of one individual fibre

Once the nodes are grouped into fibres, a linear regression is fitted for each fibre (Fig. 5).

The equation of the fitted line provides the spatial orientation of a single fibre. In addition, the center of gravity of each node cloud corresponds to the center of gravity of the fibre. Figure 6 shows a MATLAB reconstruction of the fibre orientations.

The whole process is automatized consuming an average time of three hours for each specimen in a conventional personal computer. The program requires only the next inputs:

- Number of pixels of the image, in order to process the slices.
- Slice pixel size, to assign X and Y coordinates, and also the tolerance limit for grouping the fibres.
- Distance between the different slices, to assign coordinates Z values. To ensure that the procedure has a high efficiency the distance between two sections cuts should be smaller or equal to the fibre diameter.
- Percentage fibre volume, to obtain the threshold limit.
- Theoretical length of the fibre.

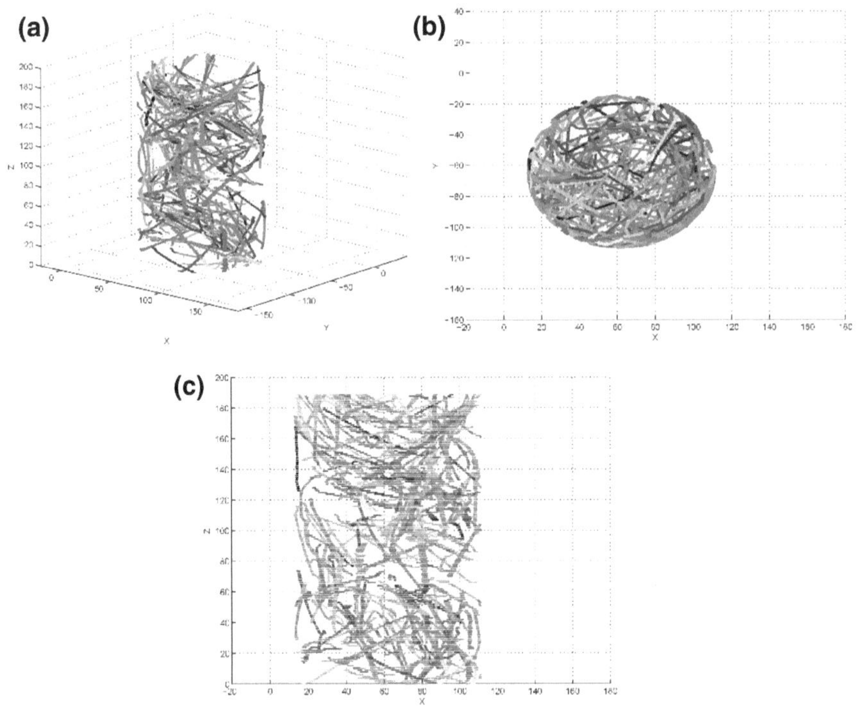

Fig. 6 MATLAB 3D reconstruction of the fibre distribution of a specimen. Isometric (**a**), XY view (**b**) and XZ view (**c**)

The flow diagram of the whole procedure is shown in Fig. 7. It is noteworthy that there is not any manual step (except for the input data), which guarantees that human errors are minimized.

The MATLAB program yields the following two data:

1. Orientation vector: $\vec{v} = \left(v_x, v_y, v_z\right)$
2. Centre of gravity: $c = \left(c_x, c_y, c_z\right)$

Using this information, the fibre orientation can be obtained and can be shown in two different coordinate systems: the global Cartesian coordinate system and the local Cartesian coordinate system (Radial–Circumferential–Vertical Coordinate System) (Fig. 8). Next, both systems are explained.

2.1 Global Cartesian Coordinate System

The fibre orientation according to the global Cartesian coordinate system is as follows:

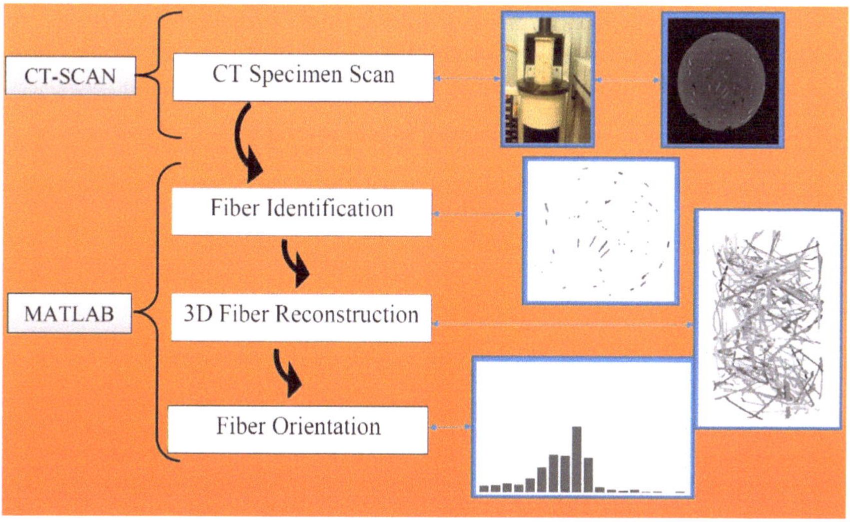

Fig. 7 Workflow

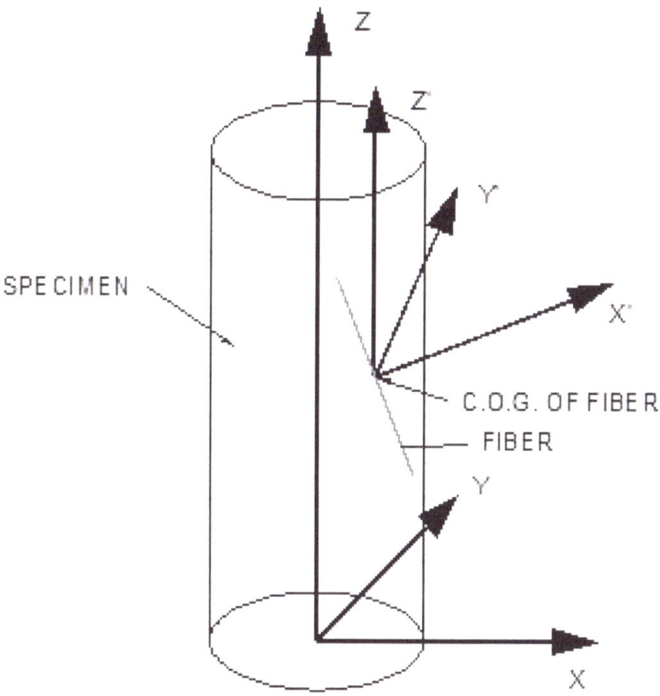

Fig. 8 Explanation of both global and local Cartesian coordinate system

1. X-axis orientation α_x : corresponding to the angle between the line and the X-axis.
2. Y-axis orientation α_y : corresponding to the angle between the line and the Y-axis.
3. Z-axis orientation α_z : corresponding to the angle between the line and the Z-axis.

The orientation of this fibre is obtained as follows (Eq. 1):

$$\cos(\alpha_x) = \frac{v_x}{\left|\vec{v}\right|}$$
$$\cos(\alpha_y) = \frac{v_y}{\left|\vec{v}\right|} \quad (1)$$
$$\cos(\alpha_z) = \frac{v_z}{\left|\vec{v}\right|}$$

Using the above mentioned information, it is possible to draw the histograms for each of the three directions X, Y and Z. In all cases, abscissa axis shows the angle and the ordinate axis shows the frequency. The results of the angles are shown in 10° ranges.

The dominant direction of the fibres corresponds to the angles that exhibit the highest frequency of occurrence.

2.2 *Local Cartesian Coordinate System*

In some applications, it may be interesting to determine the dominant direction of the fibres in a local Cartesian coordinate system, where the X'-axis corresponds to the radial direction, the Y'-axis corresponds to the circumferential direction and the Z'-axis corresponds to the vertical direction. This occurs, for example, in specimens that have outer and/or inner circular contours. Use of the local Cartesian coordinate system can show this orientation, but this phenomenon cannot be observed using a global Cartesian coordinate system.

First of all, it is necessary to define the local Cartesian coordinate system, as shown in Fig. 9.

Next, it is necessary to redefine the orientation vector $\vec{v}' = \left(v_{x'}, v_{y'}, v_{z'}\right)$ according to this coordinate system (Fig. 9) using the following equation (Eq. 2):

$$\begin{bmatrix} v_{x'} \\ v_{y'} \\ v_{z'} \end{bmatrix} = \begin{bmatrix} \cos(\alpha) & \sin(\alpha) & 0 \\ -\sin(\alpha) & \cos(\alpha) & 0 \\ 0 & 0 & 1 \end{bmatrix} \cdot \begin{bmatrix} v_x \\ v_y \\ v_z \end{bmatrix} \quad (2)$$

The fibre orientations are obtained according to the local Cartesian axes X', Y' and Z'.

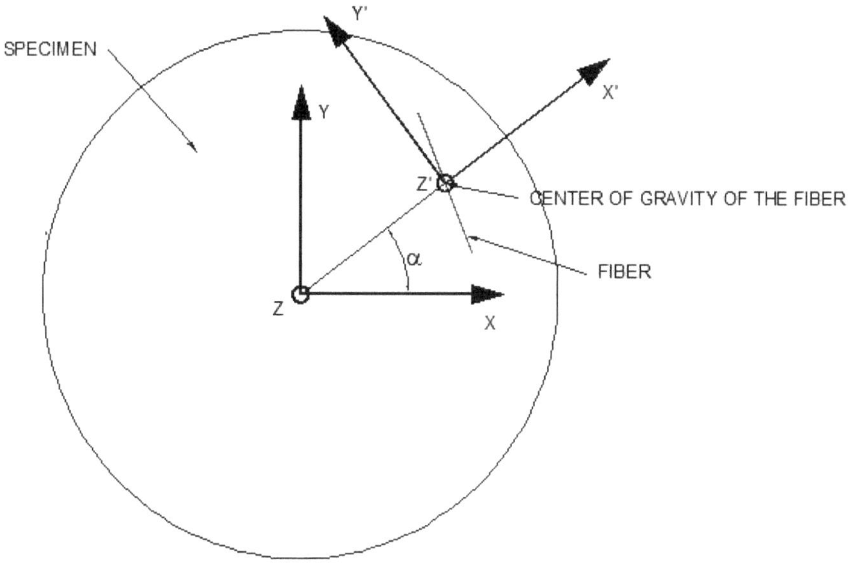

Fig. 9 Correlation between global and local Cartesian coordinate system

2.3 *Efficiency Factors*

Next, efficiency factors of the fibres in each of the three directions can be defined as follows:

1. X-axis efficiency e_x: according to the X-axis
2. Y-axis efficiency e_y: according to the Y-axis
3. Z-axis efficiency e_z: according to the Z-axis.

These values are obtained according to the following equations (Eq. 3):

$$\begin{aligned} e_x &= \sum_{i=1}^{N_b} \left(2 \cdot f_x\left(\alpha_{x,i}\right) \cdot \cos\left(\alpha_{x,i}\right)\right) \\ e_y &= \sum_{i=1}^{N_b} \left(2 \cdot f_y\left(\alpha_{y,i}\right) \cdot \cos\left(\alpha_{y,i}\right)\right) \\ e_z &= \sum_{i=1}^{N_b} \left(2 \cdot f_z\left(\alpha_{z,i}\right) \cdot \cos\left(\alpha_{z,i}\right)\right) \end{aligned} \tag{3}$$

where

N_b: number of bars of the histogram.

$f_x\left(\alpha_{x,i}\right)$: frequency of each bar of the histogram corresponding to the X-axis orientation.

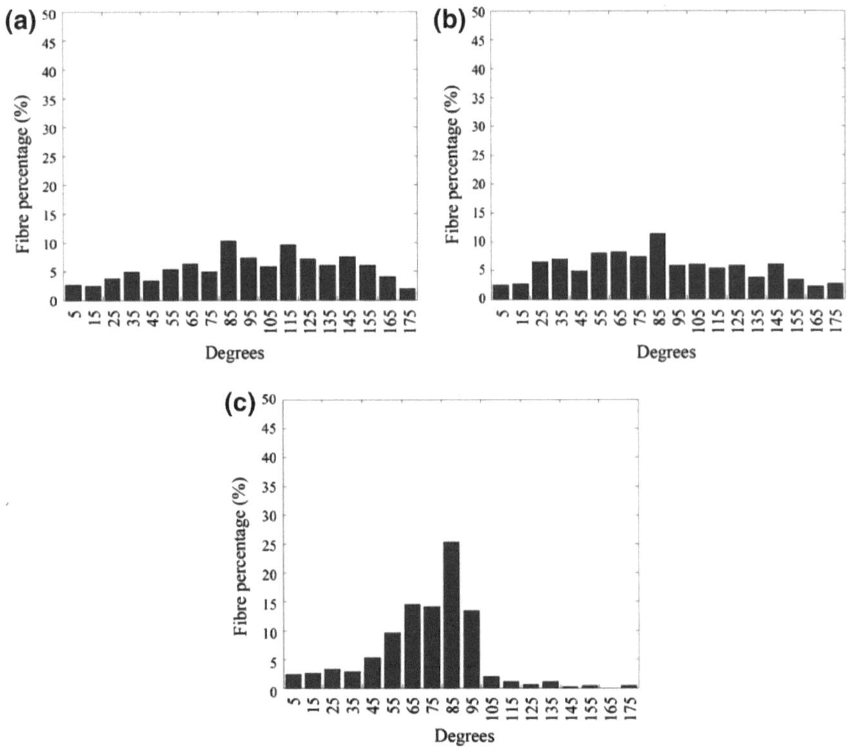

Fig. 10 Histogram X-axis (**a**), Y-axis (**b**) and Z-axis (**c**). Global Cartesian coordinate system

$f_y\left(\alpha_{y,i}\right)$: frequency of each bar of the histogram corresponding to the Y-axis orientation.

$f_z\left(\alpha_{z,i}\right)$: frequency of each bar of the histogram corresponding to the Z-axis orientation.

The range of the efficiency factors is between 0 and 2. A value of 2 means that the fibres are completely aligned with the corresponding axis. A value of 1 means that the fibres are not oriented to this axis at all. A value of 0 means that the fibres are orientated orthogonally to the corresponding axis.

Figures 10 and 11 show the average histograms of the specimen results.

Table 3 shows the efficiencies of the fibres in each of the three directions.

In this particular case, the results indicate that there is not a clear dominant direction because the efficiencies of the fibres are approximately 1 in all cases. It seems that the fibres show a tendency orthogonal to vertical direction (along Z axis). In XY or X'Y' coordinate planes no dominant direction is shown.

In this case, according to the local Cartesian coordinate system, it can be concluded that the contour has no influence on the fibre orientations.

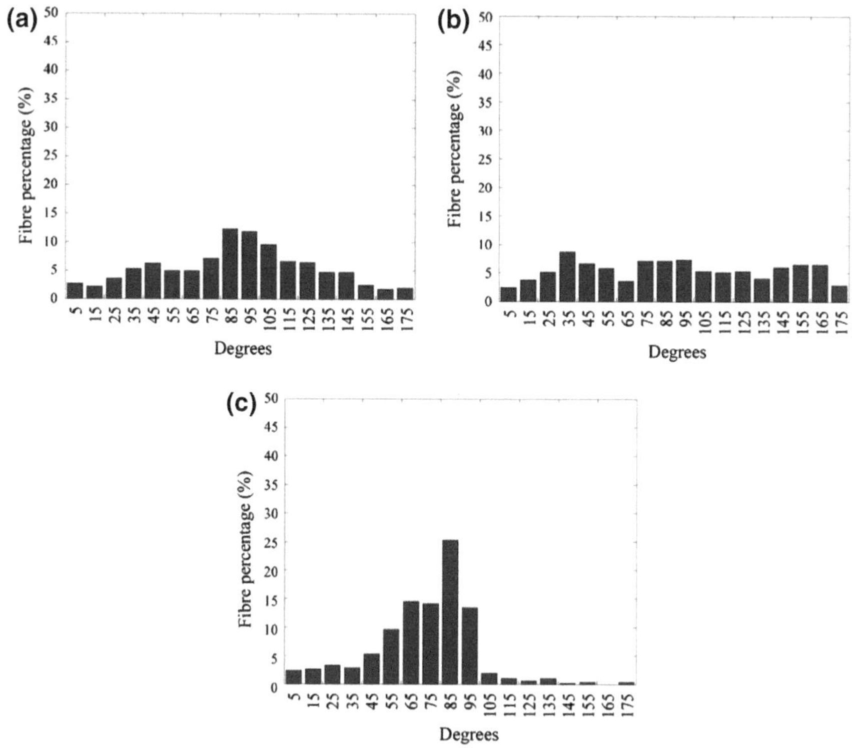

Fig. 11 Histogram X'-axis (**a**), Y'-axis (**b**) and Z'-axis (**c**). Local Cartesian coordinate system

Table 3 Efficiency of fibres

	Cartesian coordinate system			Local Cartesian coordinate system		
Sample	e_x	e_y	e_z	e_x	e_y	e_z
1	1.12	1.08	0.65	0.98	0.99	0.65
2	1.14	1.03	0.73	1.20	0.90	0.73
3	0.98	1.15	0.70	1.25	0.85	0.70
4	0.99	1.16	0.75	1.22	0.99	0.75
5	1.15	1.11	0.77	1.18	1.05	0.77
8	1.18	1.09	0.82	1.29	1.02	0.82
7	1.11	1.05	0.71	1.33	0.90	0.71
8	1.13	1.07	0.65	1.24	0.98	0.65
Average	1.10	1.09	0.72	1.21	0.96	0.72

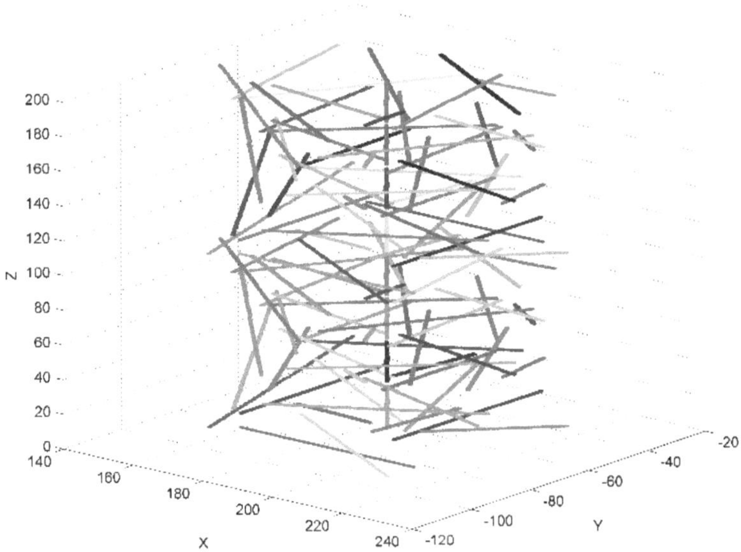

Fig. 12 Theoretical model

2.4 *Error Analysis*

Next, the error related to this procedure is analyzed. In this case, the total error of this procedure is the sum of the errors of the CT-scan and the error of the MATLAB routine.

The uncertainty related to the CT-Scan process depends on equipment precision, dimension of the testing and dimension of the fibres, among others. There is an inverse relationship between image resolution, fibre diameter and error. It is suggested to use the following criteria to guarantee an acceptable quality of the results (Eq. 4).

$$\frac{\text{Fiber diameter}}{\text{pixel size}} > 3 \tag{4}$$

In this case, a "theoretical fibre reinforced specimen" has been developed. This specimen has been performed using a 3-D drawing software and next, the slices have been obtained. All of them have been exported to .bmp files and next they are analyzed using our MATLAB routine in the same way as the slices provided by the CT-Scan. A similar method has been used in [2].

In this case, the theoretical specimen contains 104 fibres. The orientation of each fibre is known. The results obtained using the MATLAB routine have been compared to the theoretical results (Figs. 12 and 13). The error is estimated using a least-squares method. In this case, the angle error obtained is $\pm 0.41°$.

The correlation coefficient between theoretical and MATLAB routine histogram is shown in Table 4.

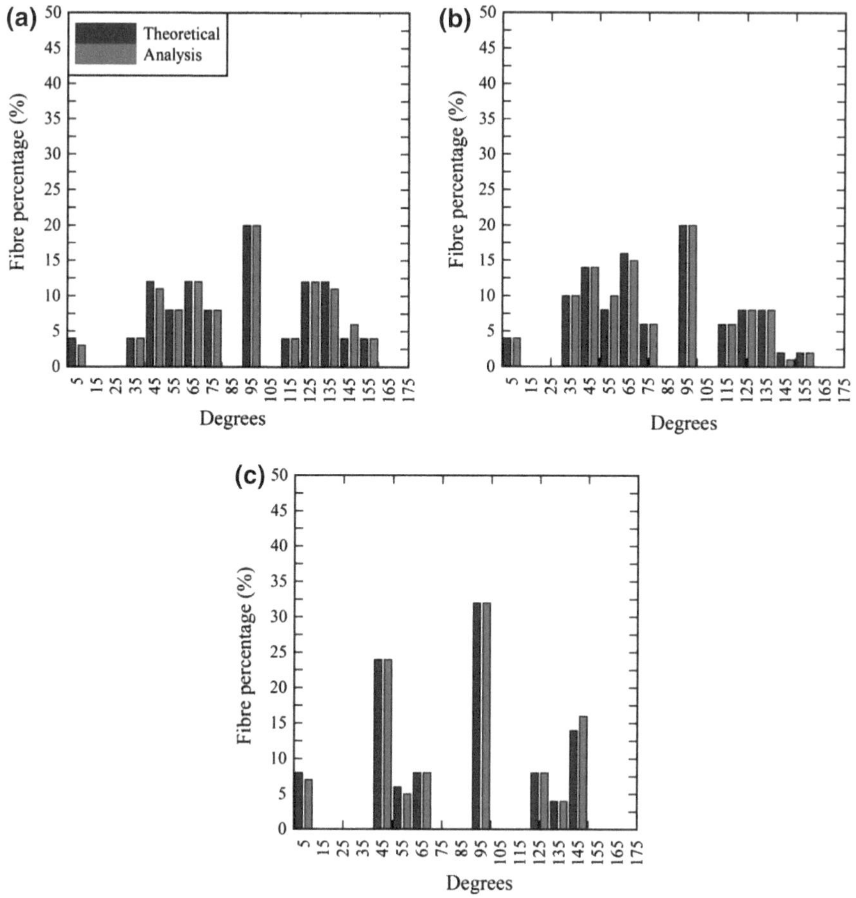

Fig. 13 Theoretical versus MATLAB Analysis. X-axis (**a**), Y-axis (**b**) and Z-axis (**c**)

Table 4 Correlation coefficient between theoretical and analysis histograms

r_x	r_y	r_z
0.9985	0.9988	0.9995

3 Summary and Conclusions

This paper shows a novel technique, implemented in a MATLAB routine, to obtain the fibre orientation in concrete specimens using CT-scan technology. This is a technology with a great potential in concrete technology because it is non-destructive and, in consequence, the specimens are not damaged.

Up to date, the main limitation in the use of this technology is the lack of post-processing subroutines, able to properly manage the huge amount of data provided by the CT-Scan.

The procedure shown in this paper is totally automatized and easy to use. It provides the individual orientation of the fibres and their center of gravity. Using this information, it is possible to obtain the dominant fibre orientation.

In this case, two coordinate systems have been performed: a global Cartesian coordinate system and a local Cartesian coordinate system. The first system provides information regarding the fibre orientations along the X, Y and Z directions. The second one provides information about the fibre orientations along the radial, circumferential and Z directions. The second coordinate system is useful in the case of concrete specimens with circular formwork (as occurs with cylindrical specimens).

This technology could help the concrete industry in developing concrete casting processes that lead to obtain proper fibre orientations inside concrete elements.

References

1. Damme, S.V., Franchois, A., Zutter, D.D., Taerwe, L.: Nondestructive determination of the steel fiber content in concrete slabs with an open-ended coaxial probe. IEEE Trans. Geosci. Remote Sens. **42**(11), 2511–2521 (2004). https://doi.org/10.1109/TGRS.2004.837332
2. Eik, M., Herrmann, H.: Raytraced images for testing the reconstruction of fibre orientation distributions. Proc. Est. Acad. Sci. **61**, 128–136 (2012). https://doi.org/10.3176/proc.2012.2.05
3. Faifer, M., Ottoboni, R., Toscani, S., Ferrara, L.R.F.: A multielectrode measurement system for steel fiber reinforced concrete materials monitoring. In: Proceedings of IEEE Instrumentation Measurement Technology Conference, pp. 313–318. Singapore (2009)
4. Faifer, M., Ottoboni, R., Toscani, S., Ferrara, L.: Nondestructive testing of steel fiber reinforced concrete using a magnetic approach. IEEE Trans. Instrum. Meas. **60**(5), 1709–1717 (2011)
5. Herrmann, H., Pastorelli, E., Kallonen, A., Suuronen, J.P.: Methods for fibre orientation analysis of X-ray tomography images of steel fibre reinforced concrete (SFRC). J. Mater. Sci. **51**(8), 3772–3783 (2016). https://doi.org/10.1007/s10853-015-9695-4
6. Jianming, G., Wie, S., Keiji, M.: Mechanical properties of steel fiber reinforced high strength, lightweight concrete. Cement Concr. Compos. **19**(4), 307–313 (1997)
7. Kang, S., Park, J., Ryu, G., Kim, S.: Investigation of fiber alignment of UHSFRC in flexural members. In: Proceedings of 8th International Symposium on Utilization of High-Strength and High-Performance Concrete, pp. 709–714. Tokyo, Japan (2008)
8. Kang, S., Lee, B., Park, Y., Kim, J.: Tensile fracture properties of an ultra high performance fiber reinforced concrete (UHPFRC) with steel fiber. Compos. Struct. **92**(1), 61–71 (2010)
9. Kang, S., Lee, B., Kim, J., Kim, Y.Y.: The effect of fiber distribution characteristics on the flexural strength of steel fiber reinforced ultra high strength concrete. Constr. Build. Mater. **25**(5), 2450–2457 (2011)
10. Kaufmann, J., Frech, K., Schuetz, P., Münch, B.: Rebund and orientation of fibers in wet sprayed concrete application. Constr. Build. Mater. **49**, 15–22 (2013)
11. Kim, J., Kim, J., Ha, G., Kim, Y.: Tensile and fiber dispersion performance of ECC (engineered cementitious composites) produced with ground granulated blast furnace slag. Cement Concr. Res. **37**(7), 1096–1105 (2007)
12. Krause, M., Hausherr, J., Burgeth, B., Herrmann, C., Krenkel, W.: Determination of the fibre orientation in composites using the structure tensor and local X-ray transform. J. Mater. Sci. **45**(4), 888–896 (2010). https://doi.org/10.1007/s10853-009-4016-4

13. Mangat, P.: Tensile strength of steel fiber reinforced concrete. Cement Concr. Res. **6**(2), 245–252 (1976)
14. Ozyurt, N., Mason, T.O., Shah, S.P.: Nondestructive monitoring of fiber orientation using AC-IS: an industrial-scale application. Cement Concr. Res. **36**(9), 1653–1660 (2006a)
15. Ozyurt, N., Woo, L., Mason, T.O., Shah, S.P.: Monitoring fiber dispersion in fiber reinforced cementitious materials: comparison of AC impedance spectroscopy and image analysis. ACI Mater. J. **103**(5), 340–347 (2006b)
16. Pastorelli, E., Herrmann, H.: Time-efficient automated analysis for fibre orientations in steel fibre reinforced concrete. Proc. Est. Acad. Sci. **65**(1), 28–36 (2016). https://doi.org/10.3176/proc.2016.1.02
17. Ponikiewski, T., Katzer, J., Bugdol, M., Rudzki, M.: Steel fibre spacing in self-compacting concrete precast walls by X-ray computed tomography. Mater. Struct. **48**(12), 3863–3874 (2015a). https://doi.org/10.1617/s11527-014-0444-y
18. Ponikiewski, T., Katzer, J., Bugdol, M., Rudzki, M.: X-ray computed tomography harnessed to determine 3D spacing of steel fibres in self compacting concrete (SCC) slabs. Constr. Build. Mater. **74**, 102–108 (2015b). https://doi.org/10.1016/j.conbuildmat.2014.10.024
19. Schnell, J., Schladitz, K., Schuler, F.: Richtungsanalyse von fasern in betonen auf basis der computer-tomographie. Beton- und Stahlbetonbau **105**(2), 72–77 (2010). https://doi.org/10.1002/best.200900055
20. Song, P., Hwang, S.: Mechanical properties of high strength steel fiber reinforced concrete. Constr. Build. Mater. **18**(9), 669–673 (2004)
21. Stroeven, P., Hu, J.: Review paper—stereology: historical perspective and applicability to concrete technology. Mater. Struct. **39**(1), 127–135 (2006). https://doi.org/10.1617/s11527-005-9031-6
22. Suuronen, J.P., Kallonen, A., Eik, M., Puttonen, J., Serimaa, R., Herrmann, H.: Analysis of short fibres orientation in steel fibre reinforced concrete (SFRC) using X-ray tomography. J. Mater. Sci. **48**(3), 1358–1367 (2013). https://doi.org/10.1007/s10853-012-6882-4
23. Torrents, J., Mason, T., Peled, A., Shah, S., Garboczi, E.: Analysis of the impedance spectra of short conductive fiber-reinforced composites. J. Mater. Sci. **36**(16), 4003–4012 (2001)
24. Vicente, M., Minguez, J., González, D.: The use of computed tomography to explore the microstructure of materials in civil engineering: from rocks to concrete. In: Halefoglu, D.A.M. (ed.) Computed Tomography-Advanced Applications. InTech (2017). https://doi.org/10.5772/intechopen.69245
25. Woo, L., Wansom, S., Hixson, A., Campo, M.A., Mason, T.O.: A universal equivalent circuit model for the impedance response of composites. J. Mater. Sci. **38**(10), 2265–2270 (2003)
26. Woo, L., Wansom, S., Ozyurt, N., Mu, B., Shah, S., Mason, T.O.: Characterizing fiber dispersion in cement composites using ac-impedance spectrometry. Cement Concr. Compos. **27**(6), 627–636 (2005)
27. Yazici, S., Inan, G., Tabak, V.: Effect of aspect ratio and volume fraction of steel fiber on the mechanical properties of SFRC. Constr. Build. Mater. **21**(6), 1250–1253 (2007)
28. Žirgulis, G., Švec, O., Geiker, M.R., Cwirzen, A., Kanstad, T.: Influence of reinforcing bar layout on fibre orientation and distribution in slabs cast from fibre-reinforced self-compacting concrete (FRSCC). Struct. Concr. **17**(2), 245–256 (2016). https://doi.org/10.1002/suco.201500064

Tunneling—Percolation Behavior of Graphene-Encapsulated Whiskers as Electroconductive Fillers for Ceramics

Irina Hussainova, Roman Ivanov, S. Sudhir Kale and Iwona Jasiuk

Abstract A direct single-step catalyst-free CVD technique has been used for producing alumina nano-whiskers covered by a few layers of defective graphene. The hybrid whiskers have been then exploited as electroconductive fillers to oxide ceramics. The electrically conductive additives do not substantially change the mechanical properties. However, the resistivity of the composites undergoes a considerable drop turning the dielectric oxides into conductive composites by addition of 2 vol% of fillers. Three-dimensional Monte Carlo simulation of systems of polydisperse prolate ellipsoids, using the critical path based tunneling-percolation model, has been exploiting for estimation of a tunnelling length-scale. The value of percolation threshold is found to be 2.23 nm for the materials under consideration, with is in a good agreement with experimental data.

1 Introduction

Miniaturization of ceramic components with complex shapes is a great challenge in many applications. The main limiting factors are mechanical unreliability and poor electrical conductivity, which make the materials difficult to be processed. Electro-discharge machining allows producing complex-shaped parts but requires the use of materials with sufficient electrical conductivity. Particular interest is currently paid to fillers of a high aspect ratio and decorated by conductive nano-species for adding required functionalities. Graphene-based additives are widely used for production of nanocomposites with an expectation to impart outstanding mechanical, electrical and thermal properties to the bulk host material. Moreover, graphene has received a great deal of attention as a possible filler for dielectric ceramics in order to turn

I. Hussainova (✉) · R. Ivanov
Tallinn University of Technology, Ehitajate 5, 19086 Tallinn, Estonia
e-mail: irina.hussainova@ttu.ee

S. S. Kale · I. Jasiuk
University of Illinois at Urbana-Champaign, Urbana, IL, USA
e-mail: ijasiuk@illinois.edu

H. Herrmann and J. Schnell (eds.), *Short Fibre Reinforced Cementitious Composites and Ceramics*, Advanced Structured Materials 95,
https://doi.org/10.1007/978-3-030-00868-0_9

them into the electroconductive state. The combination of mechanical stiffness of ceramics whiskers with electroconductivity of the deposited graphene can result in a hybrid structure with the unique combination of properties [1].

Utilization of various nanocarbon structures requires a procedure for preparation of different objects depending on specific needs. Recent studies on the growth of single-walled carbon nanotubes on oxide substrates have proved the suggestion that oxides can graphitize carbon 1. Among other methods, the chemical vapor deposition (CVD) has been demonstrated as a suitable process for the cost effective synthesis of a metal-catalyzed growth of the large-area graphene sheets. Although great efforts have been applied to explore the direct growth of graphene on dielectric substrates, the poor properties of the product produced with the help of a high-temperature synthesis or a high-cost plasma enhanced chemical vapor deposition make the catalyst-free method challenging [1, 2].

Several approaches have been proposed to synthesize graphene directly on insulating substrates without transition metals as catalysts; however, all of them required cost and time-consuming procedures. Our approach involves a single-step catalyst-free CVD for fabrication of alumina nano-whiskers encapsulated by 1–3 layers of graphene for further their application as electroconductive fillers to oxide ceramics. A range of reinforcing and toughening agents such as carbon fibers, carbon nanotubes and, recently, graphene nanoplatelets have been incorporated into ceramic matrices in an attempt to produce damage tolerant ceramic-matrix composites of suitable conductivity.

Ceramic fibers and/or whiskers have found many applications as reinforcements in composites for enhanced thermal and mechanical properties. Among a wide variety of whiskers available on the market, alumina nanofibers have received particular attention due to unique high-temperature properties and chemical stability.

The electrical conductivity is strongly dependent on an inter-filler separation. Due to the wide range of possible applications, electroconductivity of ceramics is under an intense consideration over the past decades as an in-depth understanding of the various factors governing the electrical behavior is a crucial step in the advancement of nanocomposites based technology. Generally, the conductivity of otherwise insulating ceramics can be increased by several orders of magnitude by addition of highly conductive fillers through the formation of a percolating network of electrically interconnected additives via electron tunneling [3, 4]. Because of great difference in the resistivity of the fillers and the matrix, the effective conductivity of the composite can be modeled as a tunneling-percolation problem with conductive fillers distributed in a continuous non-conductive matrix [4]. The tunneling conductance varies exponentially with the interparticle distance and the effective conductivity of the whiskers network can roughly be estimated via the critical path approximation, which allows geometrical consideration of the conductivity.

We propose a filler for production of electroconductive hierarchically structured oxide ceramics by SPS technique. Analytical studies on polydisperse rods, based on the mean-field assumption [5], have demonstrated strong sensitivity of the percolation threshold to polydispersity and aspect ratio [6]. We report on oxide ceramics added by homogeneously distributed nanofillers of a relative uniformity in length and high aspect ratio. The resistivity of the composites has undergone a considerable drop turning dielectric oxides into conductive composites by adding about 2 vol-% of nano-whiskers, which can be approximated by the prolate ellipsoids.

2 Experimental

As a matrix for composites, commercially available α-alumina, Al2O3, (Taimei TMDAR, Japan; particle size of 100 nm) powder was used. The alumina nano-whiskers of aspect ratio of 103 and a single fiber diameter of 10 ± 2 nm were fabricated with the help of chemical vapour deposition (CVD) process as detailed in [2, 7]. After 40 min of treatment in methane, which was used as a source of carbon, the mass gain of the graphene encapsulated whiskers was about 17%. Control of the deposition parameters was performed by adjustment of the ratio between a carbon source and hydrogen.

The oxide ceramics nano-powders were added by 1–7 vol% of fillers (C) and dispersed in ethanol by a stick ultrasound (Hielscher UP400S) following the procedure outlined in [8]. The mixture suspensions were treated in an attrition mill with zirconia balls of 3 mm in diameter for 1 h and then dried and sieved in 100 mm sieve. The composites were densified using a spark plasma sintering technique (Dr. SinterSPS-510CE, Japan) at 1400 °C under 40 MPa pressure for 10 min.

The microstructure was examined by scanning electron microscopy (SEM, Zeiss Gemini Ultra-55) and high-resolution transmission electron microscopy (HRTEM, JEOL 2200-S, Japan). Raman spectra were obtained with Raman spectroscopy (Horiba Jobin Yvon LabRAM 300) at laser excitation energy of 514 nm.

The density was measured by a conventional geometric method. The Vickers hardness and indentation fracture toughness (IFT) were determined using the Vickers hardness tester Indentec 5030 SKV. The IFT was determined by the measurement of the crack lengths produced by the indent following the procedure proposed by Niihara for the Palmqvist cracks [9]. The reported numbers represent the averaged values of at least ten indentations.

The room temperature electrical conductivity was measured on the prismatic bars ($25 \times 6 \times 5$ mm) machined from the original sintered disk using the 4-probe DC method with a potentiostat/galvanostat (Autolab PGSTAT 302N). Four platinum wires were attached to the specimen with a silver paste (Electrolube ERSCP03B) and connected to the terminals of the device. The electrical resistance was calculated from the linear fit of the current–voltage data points obtained from the galvanostatic tests in the range 1–10 mA with a step of 1 mA.

3 Results and Discussion

The SEM image of the precursor powder and the TEM image of the nano-whiskers are presented in Fig. 1a, b. Carbon deposited onto the surface of the nanofiber is rolled over the fibers longitudinal axis with some admixture of the opened graphene-like layers. Raman spectroscopy has revealed a well-pronounced D-peak intensity at 1337 cm^{-1} suggesting a large amount of the defective graphene sheets as this peak is related to the breathing modes of the six-atom rings and requires a defect for its activation [10], Fig. 1c. The similar Raman spectra have been obtained for damaged graphene [10] or multiwalled carbon nanotubes [10, 11]. Three most pronounced features are the G-peak at 1580 cm^{-1}, D-peak at around 1350 cm^{-1}, and 2D (or G') band at 2672 cm^{-1}. The G band, related to the C–C bond stretching, is the main Raman signature for all sp2 carbons or graphene.

Table 1 lists the basic materials properties such as the density, Vickers hardness, and indentation fracture toughness (IFT) as well as electroconductivity of the reference alumina and the materials added by the different amount of electroconductive nanofillers. Incorporation of graphenated fillers into ceramic matrix does not result in significant changes in hardness, but positively affects toughness reflected by an increase in IFT.

Evolution of the electrical conductivity of the composites as a function of fillers load is demonstrated in Table 1. Pure alumina may be considered as an isolator.

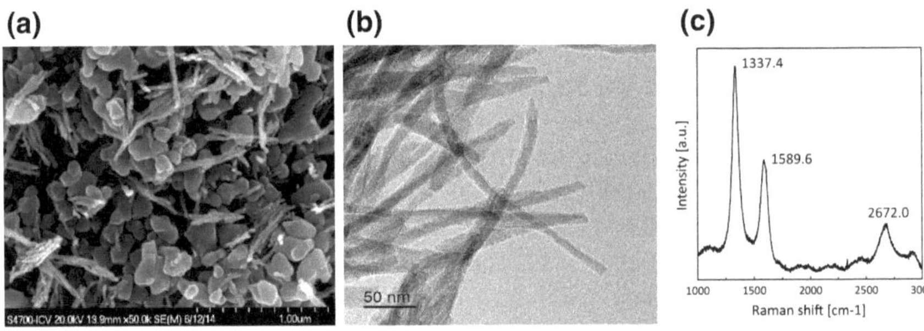

Fig. 1 FSEM images of the powder of alumina and whiskers (**a**), TEM image of the graphenated whiskers (**b**), and Raman spectrum of the graphenated nanostructures (**c**)

Table 1 Mechanical and electrical properties of materials under consideration

Material	Fillers, [vol.%]	Relative density, [%]	Hardness, [GPa]	IFT, [MPa m-1/2]	Conductivity, [S m-1]
Al2O3	0	99.6	19.1 ± 3.3	2.94 ± 0.5	0
Al2O3 – 1C	1	99.9	19.5 ± 2.1	3.87 ± 1.3	1.7 × 10 – 12
Al2O3 – 2C	2	99.8	19.75 ± 3.0	4.63 ± 0.7	10
Al2O3 – 3C	3	99.7	20.3 ± 1.5	4.96 ± 0.4	14

The addition of 1 vol-% of graphenated fillers does not result in any detectable electroconductivity. The remarkable boost in conductivity of 13 orders of magnitude was measured for the composite with 2 vol-% fillers indicating a percolation threshold for intermediate nanofillers fraction. The electrical conductivity increases sharply as the content of fillers is close to a percolation threshold, ϕ_c, which is found to be around 2 vol-% of fibers for the material under consideration. This amount of fillers corresponds to only 0.2 vol-% of carbon content in alumina. The fillers are electrically interconnected by quantum tunneling across the graphene sheets wrapped around the fibers and electron tunneling is the major charge transport mechanism in such composites3.

An effective electrical conductivity of composites can be estimated from the global tunneling network (GTN) of the conductive fillers. The tunneling conductibility diverges exponentially with the inter-fillers distance and the effective conductivity of the GTN can be predicted using the critical path approximation. The abrupt increase in dc conductivity of the ceramics beyond ϕ_c can be treated as a continuum percolation problem. Given the significant difference between the resistivity of graphene encapsulated fillers and matrix material, the focus is set only on the network of fillers. Moreover, the fillers can be modeled simply as hard-core impenetrable prolate ellipsoids.

Instead of finding the effective conductivity by numerically decimating the GTN, the conductivity can be geometrically estimated 4 as

$$\sigma \approx \sigma_0 e^{-2\frac{\delta_c}{\xi}}, \tag{1}$$

where $\delta/2$ is the thickness of the soft penetrable shell surrounding the hard particles and δ_c is the shortest distance between particles, for which the fillers at a given volume fraction form a percolating network.

The hard-core nature of fillers ensures that the fillers do not intersect each other in the model. Therefore, the condition for any two fillers to be conduct electricity, i.e., the 'connectedness criterion', should be defined. For the large fillers, i.e., when the smallest filler dimension is larger than a few micrometers, the electrical connectedness criterion is equivalent to physical contact between two fillers. However, when the smallest filler dimension is of the order of a few nanometers, electron tunneling becomes an important charge carrying mechanism. As a result, fillers that are not in direct physical contact can still be connected electrically if the separation between the fillers is in a range of a few nanometers. The tunneling effect is captured in the continuum percolation model by introducing an additional length-scale λ to define the connectedness criteria. It is expected that λ is in the range of 1–10 nm and depends primarily on the filler-matrix material combination. The connectedness criterion is then defined such that any two fillers are electrically connected when the smallest separation between their surfaces is smaller than λ. It is equivalent to adding a soft-shell of constant thickness $\lambda/2$ surrounding the hard-core fillers and stating that any two fillers are electrically connected when their soft-shells overlap. Therefore, the ϕ_c of composite can be modeled as a continuum percolation problem of hard-core fillers with soft-shells of thickness $\lambda/2$. Using various analytical methods or numerical modeling techniques3, [12], the ϕ_c can be determined if all information about filler size, shapes and spatial distribution, and λ is known. However, in practice, very limited data on filler dispersion and distribution are available and moreover, λ is not known apriori.

It should be noted that λ is a model parameter and is physically related to the strength of tunneling interactions between two given fillers in the given matrix material. Therefore, the initial objective of the modeling efforts is to use experimental data to calibrate λ for a given filler-matrix combination. This λ can then be used to make predictions about the effect of various filler attributes on ϕ_c. Analytical mean-field assumption treating polydisperse rods have shown that the percolation threshold ϕ_c is extremely sensitive to polydispersity. It was found that ϕ_c is inversely proportional to the weight average of the length distribution; with increasing polydispersity, the ϕ_c decreases even if the mean length 'L' is fixed [5].

The schematic representation of the fillers is drawn in Fig. 2a, b. After powder mixture preparation, the fibers are de-agglomerated and broken into whiskers of around 1 μm in averaged length and aspect ratio of about 1000. The nanofibers used in this study can be best captured using hard-core cylindrical fillers as shown in Fig. 2b. The lengths L of the prolate fillers has a narrow distribution around 1 μm, following some distribution function depending on the processing rout. The resulting higher moments of this length distribution are well-known to affect ϕ_c [5, 12]. The polydispersity in filler length distribution is captured by allowing the hard-core cylinder length L to follow a size distribution that is experimentally observed. The first-moment of this distribution $< L >$, i.e., the mean length and the second-moment of the distribution $< L2 >$ is relevant for the model. It is assumed that all cylindrical fillers have the same diameter D. Moreover, the fillers are assumed to be uniformly dispersed and isotropically oriented within the matrix. These are reasonable assumptions since no significant filler agglomeration or preferential filler orientation is observed. However, in general, apart from the aspect ratio and filler size polydispersity, factors such as filler agglomeration and preferential alignment of fillers are also expected to affect the percolation threshold [13].

The continuum percolation problem for the current study is to find ϕ_c for a given λ and a system of polydisperse hard-core cylindrical fillers with lengths L (following a known length-distribution) and diameter D. As volume fraction *phi* of the fillers added to the matrix increases, the average size of the clusters of interconnected fillers (according to the connectedness criteria) also increases. Eventually, at the critical volume fraction ϕ_c, a sample-spanning infinite cluster of interconnectedness fillers is formed that connects opposite ends of the sample resulting in the sudden rise in the dc conductivity. To this end, the generalized connectedness percolation theory can be employed to find ϕ_c when the average cluster size diverges to infinity [12]. A Bethe lattice mapping approach 3 can also be used and gives the same results.

For the system of polydisperse cylinders with large aspect ratios (i.e., $L >> D$), the percolation threshold is given as,

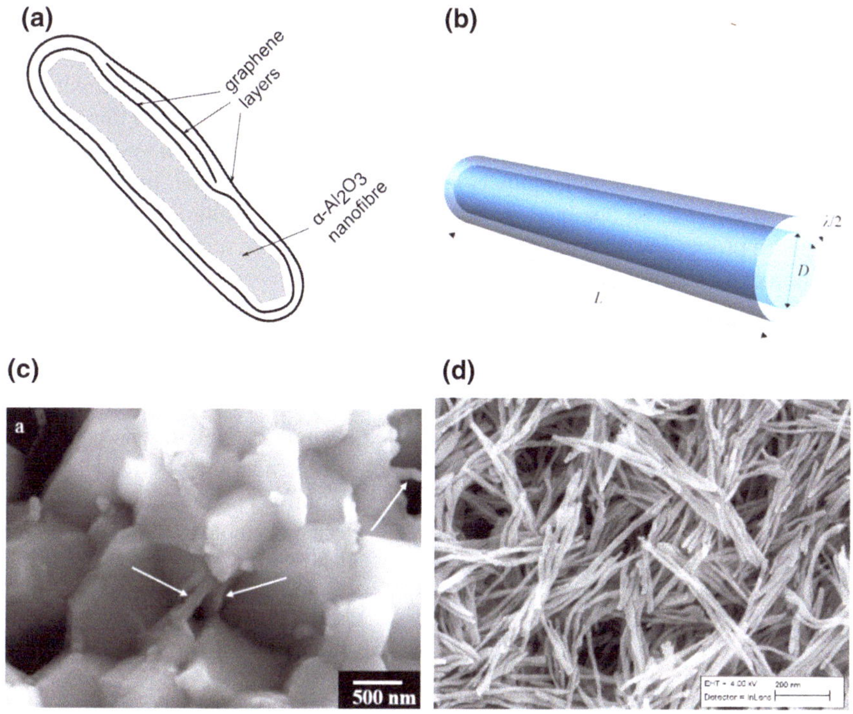

Fig. 2 Schematic representation of electroconductive filler (**a**); and schematic representation of the fillers modeled as hard-core cylinder with length L and diameter D and a soft-shell of thickness $\lambda/2$ (**b**); the fracture surface of the composite with fillers pointed by arrows (**c**); and fillers after mechanical treatment (**d**)

$$\phi_r = \frac{D^2}{2\lambda L_w} \tag{2}$$

where $L_w = \frac{<L^2>}{<L>}$ is the weight-averaged length of the nanofibers and can be calculated from fillers' length distribution. The long fibers strongly contribute to the development of a percolation network, especially in case of large difference in length for the longest and the shortest fibers. This observation is captured in Eq. (2) as ϕ_c would be smaller for a system with larger second-moment $< L2 >$ even if the mean-size $< L >$ remains unchanged. Eq. (2) has also been compared with numerical simulations of polydisperse cylindrical fillers [4, 14] and is observed to be sufficiently accurate for mean aspect ratios larger than 50. As the aspect ratios of nanofibers in this study are well beyond this number, Eq. (2) can be used for analyzing the experimental data.

In this work, the length-distribution is of a log-normal type with parameters μ and σ. Given a log-normally distributed random variable L, μ and σ are the mean and standard deviation of ln (L). For both composites considered in this study $\phi_c \approx$

2 vol-%, average length and diameter of the nanofibers after mixing are observed to be L = 1 μm and D = 10 nm, respectively. Log-normal distribution with $\sigma = 1$ is observed to well capture the size-distribution of the fillers. With these inputs and Eq. (2), the value of λ is 2.23 nm, which is within the expected range. In this way, the tunneling-percolation model of Eq. (2) was calibrated for the given system. As the polydispersity of nanofiber lengths is incorporated into the model, the quantitative estimates of ϕ_c can be obtained for any changes in filler length distribution due to changes in processing conditions and procedures. Moreover, λ obtained here can be used as a good starting assumption for making predictions on other similar systems. Considerable efforts, costs, and resources can be saved by focusing the experimental studies on volume fractions around the predicted ϕ_c from the model.

The extremely high values of electrical conductivity up to 1000 S m^{-1} have been reported for composites added by carbon nanotubes (CNT)16; however, the CNT load needed to form composites with that value is comparatively high. As a result, utilization of high graphene or CNT content leads to a considerable loss in mechanical properties of the composite as compared to the monolithic pure zirconia or alumina. Graphene augmented ceramic fibers used in this work contribute to building a percolation network, which permits decrease in required amount of carbon. Moreover, the incorporation of graphenated nanofibres renders maintenance of hardness at a high level.

4 Conclusions

Usage of graphenated electroconductive prolate fillers are feasible for fabrication of electroconductive ceramic composites, which otherwise are isolators. The low percolation threshold enables production of the ceramics without deterioration of mechanical properties but with electroconductivity that allows electro-discharge machining. Experimental data on prolate fillers are analysed using modelling based on connectedness percolation theory. Value of λ can be used as a good starting assumption for making predictions on percolation thresholds of other similar systems.

Acknowledgements The authors would like to thank the Estonian Research Council under the personal research grant PUT1063 (I. Hussainova) and the Baltic-American Freedom Foundation (BAFF) under research grant to I. Hussainova.

References

1. Ivanov, R., Hussainova, I., Aghayan, M., Drozdova, M., Perez-Coll, D., Rodriguez, M., Rubio-Marcos, F.: Graphene-encapsulated oxide nanofibers as a novel type of nanofillers for electroconductive ceramics. Eur. Ceram. Soc. (2015)
2. Ivanov, R., Mikli, V., Kübarsepp, J., Hussainova, I.: Direct CVD growth of foliated graphene closed shells on alumina nanofibers. Key Eng. Mater. (2015)

3. Ambrosetti, G., Grimaldi, C., Balberg, I., Maeder, T., Danani, A., Ryser, P.: Solution of the tunneling-percolation problem in the nanocomposite regime. Phys. Rev. B (2010)
4. Kale, S., Sabet, F., Jasiuk, I., Ostoja-Starzewski, M.: Tunneling-percolation behavior of polydisperse prolate and oblate ellipsoids. Appl. Phys. (2015)
5. Chatterjee, A.P.: Connectedness percolation in polydisperse rod systems: a modified Bethe lattice approach. Chem. Phys. (2010)
6. Kyrylyuk, A.V., van der Schoot, P.: Proc. Natl. Acad. Sci. U.S.A. (2008)
7. Hussainova, I., Drozdova, M., Aghayan, M., Ivanov, R., Pérez-Coll, D.: Graphene covered alumina nanofibers as toughening agent in alumina ceramics. Adv. Sci. Technol. (2014)
8. Drozdova, M., Hussainova, I., Pérez-Coll, D., Aghayan, M., Ivanov, R., Rodríguez, M.: A novel approach to electroconductive ceramics filled by graphene covered nanofibers. Mater. Des. (2016)
9. Niihara, K., Morena, R., Hasselman, D.P.H.: Evaluation of Klc of brittle solids by the indentation method with low crack to indent ratios. Mater. Sci. Lett. (1982)
10. Ferrari, A., Basko, D.: Raman spectroscopy as a versatile tool for studying the properties of graphene. Nature Nanotechnol. (2013)
11. Dresselhaus, M., Jorio, A., Hoffmann, M., Dresselhaus, G., Saito, R.: Perspectives on carbon nanotubes and graphene raman spectroscopy. Nano Lett. (2010)
12. Otten, R.H., van der Schoot, P.: Connectivity percolation of polydisperse anisotropic nanofillers. Chem. Phys. (2011)
13. Kale, S., Sabet, F., Jasiuk, I., Ostoja-Starzewski, M.: Effect of filler alignment on percolation in polymer nanocomposites using tunneling-percolation model. Appl. Phys. (2016)
14. Fan, Y., Kang, L., Zhou, W., Jiang, W., Wang, L., Kawasaki, A.: Control of doping by matrix in few-layer graphene/metal oxide composites with highly enhanced electrical conductivity. Carbon N. Y. (2015)